M & E

The primary need of ₹
Education Council (TEC)
will reflect the new methoɑɫ.ɔɡy ɑnɑ syllabus requirements. In presenting M & E TECbooks, we believe
that the careful collaboration between subject editors
and authors has resulted in the very best, "tailer-made"
texts which could be devised for the student — and
which are likely to be of guidance to the lecturer also.
The aim has been to provide books based on TEC's own
objectives, concisely and authoritatively presented,
priced as closely as possible to the student budget.

General Editor

Dr. Edwin Kerr
*Chief Officer, Council for National
Academic Awards*

Subject Editors

D. Anderson
*Head of Department of General
Building, Leeds College of Building*

Dr. F. Goodall
*Head of Department of Engineering,
Salford College of Technology*

L. G. North
*Head of Department of Biological Sciences,
North East Surrey College of Technology*

C. J. Thompson
*Head of Department of Science,
Matthew Boulton Technical College*

D. J. Wilks
*Head of Department of Hotel and Catering,
Middlesex Polytechnic*

G. A. Woolvet
*Head of School of Mechanical,
Aeronautical and Production Engineering,
Kingston Polytechnic*

The M&E TECbook Series

Multiple Choice Questions in
Electrical Principles
For TEC Levels I, II and III

A. DAGGER
C.Eng., Dip.Ed., M.I.E.R.E., M.B.I.M.
Senior Lecturer in Electrical Engineering,
Wigan College of Technology

Macdonald and Evans

Macdonald & Evans Ltd.
Estover, Plymouth PL6 7PZ

First published 1981

ISBN 0 7121 1274 X

Printed in Great Britain by
J.W. Arrowsmith Ltd.,
Bristol

Preface

This book consists of 360 multiple choice questions specific-ally prepared for anyone studying or teaching Electrical Principles Units at Levels I, II and III of a Technician Education Council (TEC) Certificate course in Electrical or Electronic Engineering. The Level I questions cover those electrical engineering principles objectives found in the Physical Science Units, which are taken as a prerequisite to Level II Electrical Engineering Principles in some colleges.

The questions are the result of the author's experience in the assessment of technician students at Wigan College of Technology since 1976 and almost twenty years' experience of preparing, setting, marking and validating multiple choice tests in the Royal Air Force.

I would like to express my sincere thanks to my colleague Douglas Nave, for his many helpful comments and criticisms during the preparation of this book, and to my wife Joan who spent many hours converting my almost illegible hand-writing into a readable text.

September 1981   A.D.

Contents

Introduction

Multiple choice tests, despite their extensive use over the years, are still treated with general mistrust by teachers. It is beyond the scope of this publication to discuss in detail the relative merits or otherwise of this type of testing. This has already been done very effectively by Gronlund (1965)[1], Anastasi (1968)[2], Ebel (1972)[3], Thyne[4] and others. In fact it was Ebel (1972)[3] who concluded:

> Multiple choice test items are currently the most highly regarded and widely used form of objective test items. They are adaptable to the measurement of most important educational outcomes − knowledge, understanding and judgment, ability to solve problems, to recommend appropriate action, to make predictions. Almost any understanding or ability that can be tested by means of any other item form − short answer, completion, true-false matching or essary − can also be tested by means of multiple choice test items.

A somewhat less academic reason for recommending the use of multiple choice tests is to solve a real assessment problem which the introduction of TEC schemes has presented to many teachers. That is, to satisfy the requirements for continuous assessment of TEC units. Most units are of sixty hours duration, and if continuous assessment is not to take up a disproportionate amount of the available teaching time, then quick and effective methods of testing that specific course objectives have been met are required.

Experience has shown that a twenty item multiple choice test can be taken by the students and marked and discussed by the teacher in a two hour teaching session. The aim of this book is to provide enough questions for five within-unit tests for each unit (100 items) and a further twenty items for use as alternative questions or in end-of-unit tests. As far as possible questions have been written which cover all the more important specific objectives of Level I, II and III Electrical Principle units.

The book should be particularly useful to students who wish to use the questions privately for practice, revision or self-analysis. Care must be taken to develop a correct technique when not working under supervision. When attempting multiple choice questions always, in the first instance, try to answer them without reference to any of the four possible answers. In this way you will be prevented from being diverted from the correct answer by good distractors which are an inherent part of all well-written multiple choice questions.

To facilitate the marking of multiple choice test papers, it

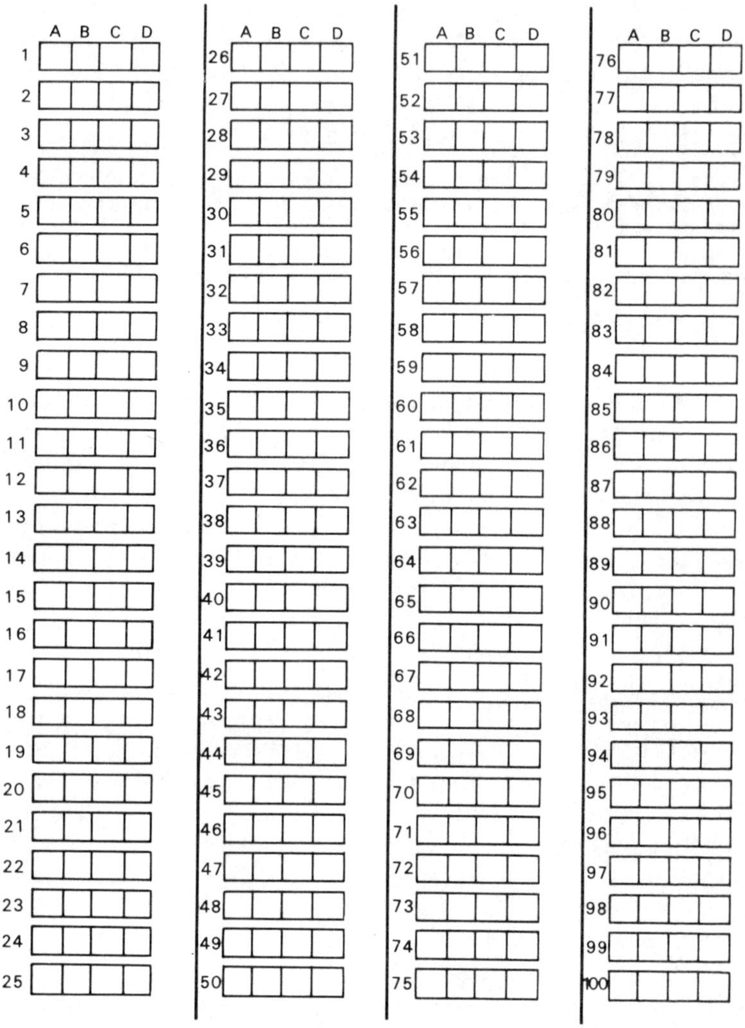

Fig. 1. *A simple type of answer grid.*

is recommended that a simple answer grid is used by the students. These answer grids can take many forms but one of the simplest and most effective designs is that illustrated in Fig. 1. Students are asked to make their responses by marking a simple cross in the appropriate box. Should a student wish to change his mind then all he has to do is to draw a circle round the incorrect response and place a new cross in the alternative box. In the unlikely event that he should wish to return to his original choice then he can indicate this by using a printers' code mark "STET" above the response which he finally wishes to be considered.

This method can be best appreciated from the illustration Fig. 2.

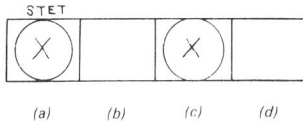

Fig. 2. *An example of a deleted then reinstated answer, where "(a)" is the required answer.*

There are on the market several sophisticated machines for marking multiple choice answer grids. All of them, however, require specially printed answer sheets or cards and careful completion by the students. They are generally expensive and would not normally be available for each teacher in the classroom. A simple but effective marking system is to punch holes in a duplicate answer grid to that shown in Fig. 1, corresponding to the correct responses. The teacher should first identify and mark any "STET" items before placing the answer grid over each student's answer grid in turn, revealing only the correct responses. A mark is then made through the appropriate aperture and a tally of the total correct responses quickly made. Using this method, a twenty item answer grid can be marked accurately in approximately thirty seconds. The biggest advantage of this simple system is that it can be carried out almost anywhere; the classroom, office, train or bus etc.

Having marked the multiple choice question papers it is important that the teacher should subject each test to a simple statistical analysis. This analysis should indicate to the teacher:

(a) the difficulty of each item;

(b) the discriminating power of each item;

(c) any errors in the item.

There are many statistical tests which can be applied, but perhaps the two simplest and most useful are the determination of the Facility (or Difficulty) Value (F.V.) and the Index of Discrimination (I.D.).

Facility (or Difficulty) Value (F.V.)

$$\text{F.V. (\%)} = \frac{\text{Total number making correct choice}}{\text{Total number of candidates}} \times 100\%$$

Although relatively easy to calculate, the interpretation of the resulting F.V. is statistically quite complex. A useful guide to the difficulty of items[5] is:

very easy	— 90 per cent and above;
easy	— 70-90 per cent;
fairly difficult	— 30-69 per cent;
difficult	— Below 30 per cent.

Adkins[6] has suggested that a figure of 62 per cent is an optimum figure for a good question. The average F.V. for all the items on a paper, depends on what is the object of the paper and the level of the candidates. It may be that there is a valid educational reason for setting a particularly difficult paper (say F.V. = 40 per cent), or a very easy paper (say F.V. = 75 per cent). This is for the individual teacher to decide. However, whenever a particular item receives an F.V. of O, then it should be very carefully examined to see whether or not the subject matter has been covered or if the question is correct.

Index of Discrimination (I.D.)

There are several different I.D.s. One of the simplest to operate in practice is that attributed to Kelley[7].

$$\text{I.D.} = \frac{\left(\begin{array}{c}\text{no. in top 27\% of}\\\text{candidates making}\\\text{correct response}\end{array}\right) - \left(\begin{array}{c}\text{no. in bottom 27\% of}\\\text{candidates making}\\\text{incorrect responses}\end{array}\right)}{27\% \text{ of total candidates}}$$

This index of discrimination has a range from -1.00 to $+1.00$. The higher the value of the index, the greater the tendency for a class doing well on an item to do well on the test paper as a whole. Again there are no hard and fast rules for selecting a specific value of index, but a useful guide is:

0.4 and above — good discrimination;
0.30 − 0.39 — reasonable discrimination;
0.2 − 0.29 — marginal discrimination;
below 0.2 — poor discrimination.

A negative index of discrimination, indicates that the correct choice has been made more frequently by "poor" candidates than by "good" candidates, and the reliability of the question should be queried and probably eliminated.

For more details of these and other statistical tests, the reader is referred to the Bibliography, Appendix II. Suffice it to say, it is appreciated that busy teachers will be hard pressed to find the time to carry out much more analysis than suggested above.

CHAPTER ONE

Electrical Principles Level I

```
CHAPTER OBJECTIVES

After studying this chapter you should be able to:

*   explain the simple ideas of atomic structures and electric
    current (1–28);
*   understand the effects of an electric current (29–50);
*   understand the concept of an electromagnetic field and
    its applications (51–73);
*   understand series and parallel resistor combinations and
    the effect on resistance of the dimensions of a conductor
    (74–96);
*   explain the effects of temperature on resistance (97–105);
*   explain the construction and principles of operation of
    moving coil and moving iron type of instruments
    (106–120).
```

The simple ideas of atomic structures and electric current

1. A simple atom consists of a
   (a) nucleus and orbital proton
   (b) nucleus and orbital electron
   (c) electron and orbital proton
   (d) proton and orbital neutron.

2. The three basic particles in an atom are
   (a) nucleus, proton, neutron
   (b) electron, ion, proton
   (c) nucleus, electron, proton
   (d) electron, proton, neutron.

3. An atom which has lost an electron is called a
   (a) positive ion
   (b) neutron
   (c) proton
   (d) negative ion.

4. A substance whose molecules contain only one kind of atom is called
   (a) a mixture
   (b) a compound
   (c) an element
   (d) an electrolyte.

5. Comparing electrons and protons it is true to say that
   (a) electrons are lighter and positively charged
   (b) electrons are lighter and negatively charged
   (c) protons are lighter and positively charged
   (d) electrons are heavier and negatively charged.

6. A copper atom consists of
   (a) protons orbiting a central nucleus
   (b) a nucleus orbiting central electrons
   (c) electrons orbiting a central proton
   (d) electrons orbiting a central nucleus.

7. An atom consists basically of
   (a) a positive nucleus with protons revolving around it
   (b) a neutral central proton surrounded by negative electrons
   (c) a positively charged nucleus with one or more electrons revolving around it
   (d) an electron with a nucleus revolving around it at high speed.

8. Which of the following is correct?
   (a) A negative charge repels a positive charge
   (b) A positive charge attracts a positive charge
   (c) A positive charge repels an uncharged body
   (d) A negative charge attracts an uncharged body.

9. An atom having a neutral electronic charge must contain as many electrons as
   (a) molecules
   (b) protons
   (c) neutrons
   (d) protons and neutrons.

10. An electric current is a movement of the electrons
    - (a) orbiting the nucleus of an atom
    - (b) between atoms
    - (c) within the nucleus of an atom
    - (d) which repel neutrons in an atom.

11. An electric current is
    - (a) a movement of electric charges
    - (b) atoms in a state of vibration
    - (c) a movement of protons or neutrons
    - (d) static electric charges.

12. The number of electrons passing a given point in an electrical circuit in one second is a measure of
    - (a) potential difference
    - (b) current
    - (c) charge
    - (d) power.

13. An insulator has
    - (a) many free electrons
    - (b) few free electrons
    - (c) many free protons
    - (d) few free protons.

14. A current of 10 mA may be rewritten as
    - (a) 1000 A
    - (b) 0.01 A
    - (c) 0.001 A
    - (d) 0.0001 A.

15. A current of 4 ampere flows in a circuit for 20 seconds. How much charge is transferred?
    - (a) 0.2 C
    - (b) 5 C
    - (c) 24 C
    - (d) 80 C.

16. Which of the currents listed is equivalent to 300 μA?
    - (a) 300×10^{-6} A
    - (b) 300×10^{-3} A
    - (c) 300×10^{3} mA
    - (d) 300×10^{-6} mA.

17. A steady flow of charge is maintained through a conductor. If 15 mC pass through in 2.5 ms, then the current is
    (a) 6 A
    (b) 37.5 A
    (c) 6 μA
    (d) 37.5 μA.

18. To change the unit of milliamps to microamps you would
    (a) multiply by 1000
    (b) multiply by 1 000 000
    (c) divide by 1000
    (d) divide by 1 000 000.

19. Before current can flow through a conductor
    (a) there must be a potential difference across its ends
    (b) the conductor must have adequate insulation
    (c) the resistance of the conductor must be low
    (d) the conductor must have no free electrons.

20. The difference in potential between two unequal charges is measured in
    (a) coulombs
    (b) joules
    (c) volts
    (d) amperes.

21. Ohm's law can be stated in terms of voltage V, current I and resistance R as follows
    (a) $I = \dfrac{R}{V}$
    (b) $V = IR$
    (c) $R = \dfrac{I}{V}$
    (d) $I = VR$.

22. The work done in moving 2 coulombs of charge from one point in an electric circuit to another is 5 joules. The potential difference between the points is
    (a) 0.4 V
    (b) 2.5 V
    (c) 10 V
    (d) 7 V.

23. The resistance of a conductor depends upon the
    (a) length, cross sectional area, temperature and material from which it is made
    (b) cross sectional area, current and temperature

(c) p.d. applied and the temperature of the conductor

(d) current flowing and the material 'from which it is made.

24. 5000 ohms can be written as
   (a) 5 kΩ
   (b) 5 mΩ
   (c) 5 MΩ
   (d) 5 Ω.

25. When the current in a 2 kilohms resistor is 50 milliamps the voltage across the resistor is
   (a) 100 V
   (b) 1 V
   (c) 100 mV
   (d) 2 V.

26. If the voltage across a 100 Ω resistor is 200 V then the current flowing through the resistor is
   (a) ½ A
   (b) 2 A
   (c) 200 mA
   (d) 0.5 mA.

27. In a simple resistive circuit fed from a direct current supply, the current will increase if
   (a) circuit resistance increases
   (b) applied voltage and resistance remain constant;
   (c) circuit resistance decreases
   (d) applied voltage decreases.

28. If the potential difference across a circuit is doubled and the value of the circuit resistance is halved then the current will
   (a) not alter
   (b) double
   (c) halve
   (d) increase by a factor of four.

The effects of an electric current

29. The three basic effects of an electric current are
    (a) heating, magnetic and lighting
    (b) chemical, magnetic and heating
    (c) magnetic, movement and chemical
    (d) movement, chemical and heating.

30. If a voltage of V volts is applied to a resistor of R ohms
    a current of I amps flows for t seconds. The electrical
    energy used is given by
    (a) IRt joules
    (b) VIt watts
    (c) $\dfrac{V^2}{R}$ amps
    (d) ItV joules.

31. A current of 10 mA flows through a 1 kilohm resistor
    for 2 minutes. The power dissipated by the resistor is:
    (a) 1200 mW
    (b) 0.1 W
    (c) 1000 mW
    (d) 120 mW.

32. What is the rate at which work is done when a current
    of 0.5 A flows for 10 seconds through a resistor of
    1000 Ω
    (a) 5 coulombs/second
    (b) 250 watts/second
    (c) 250 watts
    (d) 2 500 joules/second.

33. If the voltage across a resistor is doubled then the power
    developed in the resistor will be
    (a) halved
    (b) doubled
    (c) four times as great
    (d) eight times as great.

34. 1 kWh is equivalent to
    (a) 3 600 000 joules
    (b) 3600 joules
    (c) 1000 units
    (d) 60 000 joules.

35. The constant which relates electrical energy to the heat produced is called
    (a) electrochemical equivalent
    (b) joule's equivalent
    (c) temperature coefficient
    (d) electrothermal constant.

36. Wire-wound resistors are preferred to carbon resistors when
    (a) negative temperature coefficients are required
    (b) small physical size is required
    (c) large amounts of heat have to be dissipated
    (d) very large resistance values are required.

37. Electrical energy is measured in
    (a) watts
    (b) coulombs
    (c) joules
    (d) volts.

38. The mass of material deposited on an electrode in an electrolytic cell is determined by
    (a) voltage, electrochemical equivalent, thickness of electrode
    (b) current, electrochemical equivalent, time
    (c) voltage, current, time
    (d) resistance, electrochemical equivalent, current.

39. If the current flowing for 10 minutes through a copper sulphate electrolytic cell were doubled, the mass of copper deposited in this time would
    (a) halve
    (b) double
    (c) increase by a factor of four
    (d) remain constant.

40. A disadvantage of a primary cell when compared with a secondary cell is that it
    (a) is not portable
    (b) can supply only heavy currents
    (c) requires more maintenance
    (d) is not rechargeable.

41. Polarisation of a primary cell is caused by
    (a) electrolyte which is too weak
    (b) an accumulation of hydrogen gas on the positive plate
    (c) the formation of manganese dioxide
    (d) the reversal of the polarity of the terminals.

42. The instrument used to check the state of charge of a lead-acid cell by checking the electrolyte is
    (a) an ammeter   (c) a hydrometer
    (b) a voltmeter   (d) a wattmeter.

43. A major advantage of a secondary cell compared with a primary cell is
    (a) it contains sulphuric acid
    (b) it is readily portable
    (c) it requires no maintenance
    (d) it is rechargeable.

44. The specific gravity and voltage per cell of a fully charged lead-acid battery are
    (a) 1.86 and 2.2 V
    (b) 1.15 and 2 V
    (c) 1.25 and 2.2 V
    (d) 1.25 and 1.5 V.

45. The effective e.m.f. acting in this circuit is
    (a) 18 V
    (b) 12 V
    (c) 6 V
    (d) 24 V.

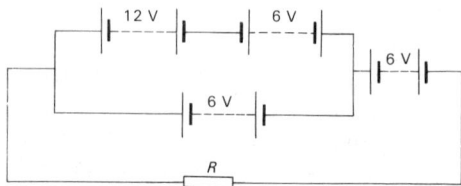

46. The internal resistance of a cell containing two plates is 0.1 Ω and the e.m.f. is 2V. If a battery consists of six 2 plate cells in series then
    (a) the battery e.m.f. is 2 V and its internal resistance 0.6 Ω
    (b) the battery e.m.f. is 6 V and its internal resistance 0.1 Ω

(c) the battery e.m.f. is 12 V and its internal resistance 0.1 Ω

(d) the battery e.m.f. is 12 V and its internal resistance 0.6 Ω.

47. If excess currents are continually drawn from lead-acid accumulators, or they are poorly maintained, then it is probable that
   (a) the internal resistance will decrease
   (b) the terminal voltage will rise
   (c) acid will leak from the cells
   (d) sulphation will occur.

48. A resistor is marked with four coloured bands. The fourth coloured band indicates
   (a) stability
   (b) tolerance
   (c) temperature coefficient
   (d) wattage rating.

49. The value of the resistor shown is
   (a) 250 ohms
   (b) 250 kilohms
   (c) 2 megohms
   (d) 520 ohms.

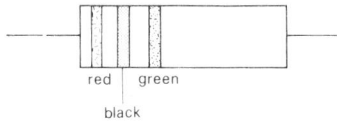

red   green

black

50. The diagram represents a colour coded resistor. Its resistance value must be between
   (a) 42.3 kΩ and 51.7 kΩ
   (b) 46.65 kΩ and 49.35 kΩ
   (c) 522 kΩ and 368 kΩ
   (d) 425.7 Ω and 520.3 Ω.

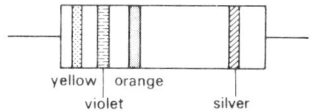

yellow   orange

violet   silver

The concept of an electromagnetic field and its applications

51. If a magnet is freely suspended in air
   (a) it comes to rest with its north pole pointing north
   (b) it continues to rotate in a clockwise direction
   (c) it comes to rest with its north pole pointing south
   (d) it continues to rotate in an anticlockwise direction.

52. The number of lines of magnetic flux passing through 1 square metre is called
    (a) magnetomotive-force
    (b) flux density
    (c) the weber
    (d) residual magnetism.

53. The magnetic pole marked X on the figure is
    (a) north
    (b) south
    (c) either north or south
    (d) neutral.

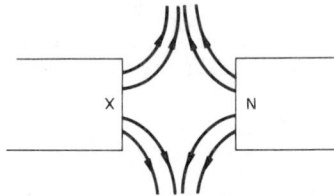

54. The force between the south pole and the piece of soft iron will be
    (a) repulsion
    (b) attraction
    (c) decreasing as d decreases
    (d) no force.

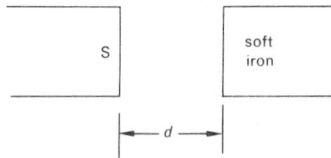

55. A plotting compass is bought close to a bar magnet and is deflected as shown on the figure. The pole marked X is
    (a) either north or south
    (b) neither north nor south
    (c) north
    (d) south.

56. The unit of flux density is
    (a) henry
    (b) weber
    (c) farad
    (d) tesla.

57. If the total flux in a magnetic circuit is 5 mWb and the cross-sectional area of the circuit is 10 square cm the flux density will be
   (a) 0.5 T
   (b) 5 T
   (c) 50 T
   (d) 50 mT.

58. This figure represents the cross section of a conductor with an electric current flowing through it. The resulting magnetic field will be
   (a) radial lines from the centre of the conductor
   (b) radial lines towards the centre of the conductor
   (c) concentric lines in a clockwise direction
   (d) concentric lines in a anticlockwise direction.

59. The flux density in a coil can be increased by
   (a) reducing the current through the coil
   (b) decreasing the number of turns of the coil
   (c) increasing the length of the coil
   (d) decreasing the cross-sectional area of the coil.

60. The force acting between the solenoid and the permanent magnet will be
   (a) one of attraction
   (b) independent of current I
   (c) zero unless the current is changing
   (d) one of repulsion.

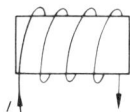

61. If the total flux in a magnetic circuit is 1mWb and the area through which it acts is 5 cm^2, the flux density in teslas is
   (a) 5×10^{-15} T   (c) 2 T
   (b) 5×10^{-7} T   (d) 2×10^{-2} T.

62. When using the Left Hand Rule, the fingers of the hand represent the directions as follows
    (a) forefinger − field, index finger − current, thumb − motion
    (b) first finger − field, second finger − current, thumb − motion
    (c) thumb − motion; first finger − current, second finger − field
    (d) thumb − magnetic field, first finger − current, middle finger − motion.

63. A magnet is moved towards a coil and comes to rest 2 cm away from it. The final induced e.m.f. is
    (a) proportional to the distance of the magnet from the coil
    (b) proportional to the square of the distance of the magnet from the coil
    (c) inversely proprotional to the square of the distance of the magnet from the coil
    (d) zero.

64. A conductor 250 mm long carries a current of 75 A and is placed at right angles to a magnetic field of density 0.8 teslas. The force on the conductor will be
    (a) 1.5 N
    (b) 25 N
    (c) 0.025 N
    (d) 15 N.

65. The effect of reversing the current in the conductor shown in the figure will be to
    (a) reduce the force F to zero
    (b) reverse the direction of the force F
    (c) increase the size of force F
    (d) have no effect on force F.

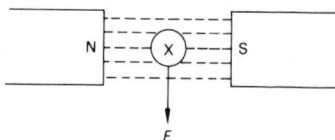

66. When a coil of wire is rotated in a magnetic field, the induced e.m.f. is proportional to
    (a) the strength of the magnetic field only
    (b) the speed of rotation only
    (c) the strength of the magnetic field, the speed of rotation and the number of turns on the coil
    (d) the number of turns on the coil only.

67. This figure represents two current-carrying solenoids. There will be
    (a) no resultant magnetic flux
    (b) a cancellation of back e.m.f.s
    (c) a force of repulsion between solenoids
    (d) large eddy current losses.

    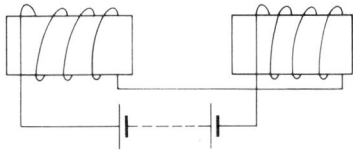

68. The magnitude of the force exerted on a conductor placed in a magnetic field depends upon
    (a) direction of flow of current through the conductor;
    (b) length, cross-sectional area of conductor and current passing through the conductor
    (c) length of conductor, current passing through conductor and flux density of magnetic field
    (d) rate of change of magnetic flux.

69. A coil of 100 turns produces a magnetic flux of 50 webers when a current of 2 amps flows in it. If this current is reduced to zero in 5 seconds the induced e.m.f. will be
    (a) 1000 V   (c) 250 V
    (b) 100 V   (d) 50 V.

70. When the magnetic flux linking a circuit changes, an e.m.f. is induced, the magnitude of which is proportional to
    (a) the square of the current passing
    (b) the maximum value of current passing
    (c) the rate of change of voltage
    (d) the rate of change of flux.

71. When the magnetic flux linking a circuit changes an e.m.f. is induced, the direction of this induced e.m.f. is such as
    (a) to assist in the build up of magnetic flux
    (b) to oppose the change producing it
    (c) to always be positive with respect to earth
    (d) to always be negative with respect to earth.

72. A conductor 500 mm long moves at 25 metres per second through a magnetic field. If the e.m.f. generated is 125 mV the magnetic field will have a flux density of
    (a) 0.01 T
    (b) 20 T
    (c) 5 T
    (d) 100 T.

73. A force of 10 N is experienced by a conductor 100 cm long as it moves through a magnetic field of density of 0.2 T. The current flowing in the conductor is
    (a) 0.5 A
    (b) 0.1 A
    (c) 10 A
    (d) 50 A.

Series and parallel resistor combinations and the effect on resistance of the dimensions of a conductor

74. The resistance between the opposite faces of a 1 metre cube of carbon material is 1 ohm. What is the resistance between the opposite faces of 1 cm cube of the same material?
    (a) 10 kilohms
    (b) 10 milliohms
    (c) 1 micro-ohm
    (d) 100 ohms.

75. The resistance of a copper conductor
    (a) varies inversely with its length
    (b) decreases as temperature increases

(c) is greater than iron of the same dimensions

(d) varies inversely with its cross-sectional area.

76. The resistivity of copper is 1.5 micro-ohm-metre. The resistance of a piece of copper wire 200 metres long and cross-sectional area 1 square mm is

(a) 0.0003 Ω

(b) 3 Ω

(c) 30 Ω

(d) 300 Ω.

77. The resistance of a wire of given length at constant temperature is

(a) independent of the material of the wire

(b) inversely proportional to the area of cross section

(c) dependent on the material and inversely proportional to the radius

(d) proportional to the area of cross section.

78. The length of wire in a wire wound resistor is increased four times and the cross sectional area of the wire is doubled. The resistance is effectively

(a) increased eight times

(b) decreased by half

(c) doubled

(d) decreased eight times.

79. The current flowing in the wire marked B in the diagram is

(a) 3 A flowing from X to Y

(b) 3 A flowing from Y to X

(c) 4 A flowing from Y to X

(d) 19 A flowing from Y to X.

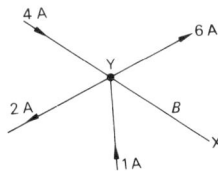

80. The current I is

(a) 5 A

(b) 19 A

(c) 3 A

(d) 8 A.

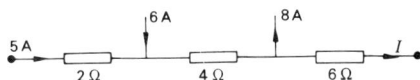

81. The value of the potential difference V_1 is
    (a) 20 V
    (b) 5 V
    (c) 10 V
    (d) 15 V.

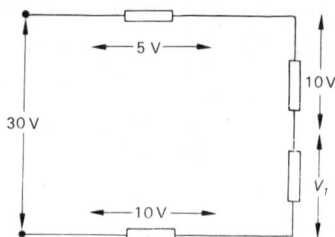

82. In this circuit the voltage across the 2 kΩ resistor is
    (a) 2 V
    (b) 4 V
    (c) 1 V
    (d) 5 V.

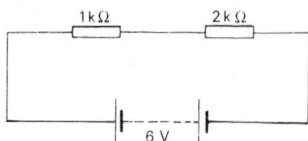

83. The voltage across the 5 kΩ resistor is 50 V, the voltage across the 3 kΩ resistor is
    (a) 30 V
    (b) 150 V
    (c) cannot be found
    (d) 15 V.

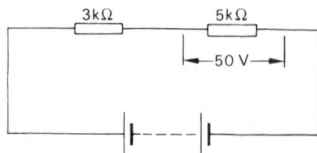

84. The equivalent resistance (R_T) of the three resistors R_1, R_2 and R_3 connected in parallel can be found from
    (a) $R_T = R_1 + R_2 + R_3$
    (b) $R_T = \dfrac{1}{R_1} + \dfrac{1}{R_2} + \dfrac{1}{R_3}$
    (c) $R_T = \dfrac{R_1 + R_2}{R_1 \times R_2} + R_3$
    (d) $\dfrac{1}{R_T} = \dfrac{1}{R_1} + \dfrac{1}{R_2} + \dfrac{1}{R_3}$

85. What is the internal resistance of a cell of e.m.f. 20 V if it has a terminal p.d. of 18 V when connected to a 36 ohm load?
    (a) 4 ohm   (c) 1 ohm
    (b) 2 ohm   (d) 0.5 ohm.

86. The voltage across the 4 Ω resistor is
   (a) 6 V
   (b) 4 V
   (c) 2 V
   (d) 8 V.

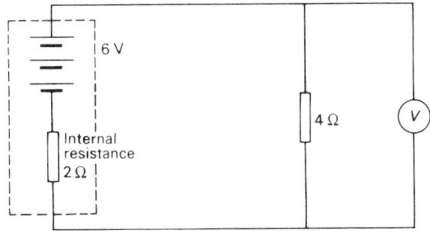

87. In the circuit shown
   (a) $I_1 R_2 = I_2 R_1$

   (b) $I_1 = \dfrac{I_2 R_2}{R_1}$

   (c) $I_1 = \dfrac{IR_1}{R_1 + R_2}$

   (d) $I = \dfrac{IR_2}{R_1 + R_2}$

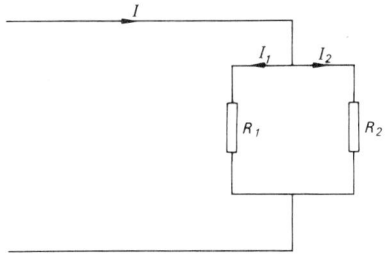

88. When the switch is closed
   (a) I_1 rises
   (b) I_2 falls
   (c) I_1 remains constant
   (d) I_1 falls.

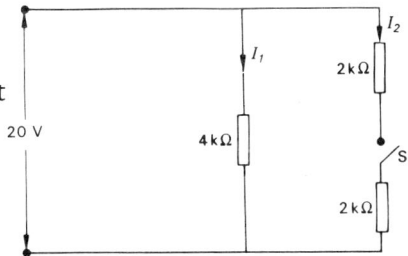

89. The supply voltage V is
   (a) 20 V
   (b) 40 V
   (c) 32 V
   (d) 24 V

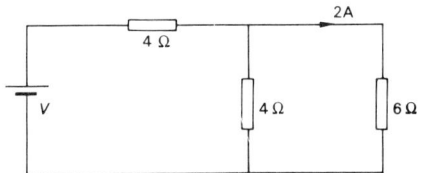

90. The variable resistor R in the figure can be varied from
    0 to 10 Ω. The resistance between A and B can vary
    between
    (a) 10 Ω and 20 Ω
    (b) 5 Ω and 10 Ω
    (c) 10 Ω and 15 Ω
    (d) 0 and 3.3 Ω (approx.).

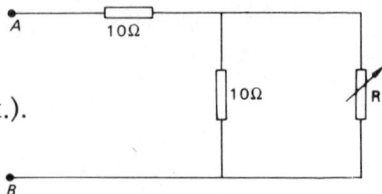

91. In this circuit the current taken from the battery is 5 A.
    The value of R is
    (a) 25 Ω
    (b) 50 Ω
    (c) 100 Ω
    (d) 200 Ω.

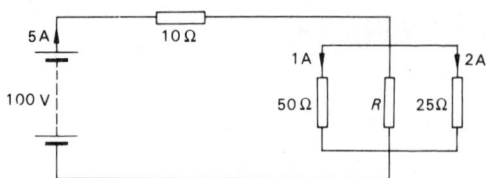

92. A variable resistor R is varied from 2 ohms to 0 ohms.
    What is the variation in current drawn from the supply?
    (a) 0 A to 3 A
    (b) 0 A to 4 A
    (c) 4 A to 6 A
    (d) 3 A to 6 A.

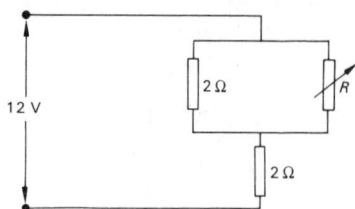

93. In the circuit shown the potential difference between
    point C and E is
    (a) 3 V
    (b) 10 V
    (c) 15 V
    (d) 13 V.

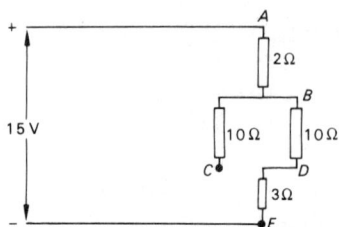

94. In the circuit shown $V_1 = V_2$. The resistance R must be
    (a) 7.5 Ω
    (b) 10 Ω
    (c) 15 Ω
    (d) 20 Ω.

95. The resistance between terminals A and B in the circuit shown is

    (a) $R_1 + \dfrac{R_2 R_3}{R_2 + R_3}$

    (b) $R_1 + \dfrac{1}{R_2} + \dfrac{1}{R_3}$

    (c) $\dfrac{R_1 R_2}{R_1 + R_2} + R_3$

    (d) $R_1 + \dfrac{R_2 + R_3}{R_2 R_3}$.

96. In the circuit shown the resistance R is reduced, therefore
    (a) I rises
    (b) I falls
    (c) I_1 remains constant
    (d) I_2 falls.

The effects of temperature on resistance

97. The power dissipated in a 10 Ω resistor is 1000 W. The current flowing in the resistor is
    (a) 100 A
    (b) 10 A
    (c) 10 000 A
    (d) 0.01 A.

98. The voltage across the terminals of an open circuited cell is 1.5 V. When supplying a 2 ohm load the voltage falls to 1 V. The power developed in the internal resistance is
    (a) 0.25 watts
    (b) 1 watt
    (c) 4 watts
    (d) 0.75 watts

99. A resistor capable of high heat dissipation is usually
    (a) made of cracked carbon
    (b) made of vitreous enamel
    (c) made of carbon
    (d) wire wound on a ceramic former.

100. A watt is the rate of working when electrical energy is
     transformed to another form of energy at the rate of
    (a) one joule per second
    (b) one coulomb per second
    (c) one newton per second
    (d) 3.6 joules per second.

101. In order to double the power dissipated in R_2, E must
     be
    (a) doubled
    (b) increased by a factor
        of $\sqrt{2}$
    (c) increased by 50%
    (d) increased by 25%.

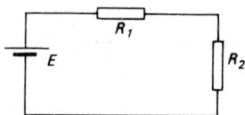

102. The resistance of most metallic conductors:
    (a) decreases with increase in temperature;
    (b) is independent of increase in temperature;
    (c) increases with increase in temperature;
    (d) is inversely proportional to temperature.

103. This resistance/temperature graph shows that the
     material under test has a
    (a) positive temperature   (c) high resistance
        coefficient   (d) negligible resistance.
    (b) negative temperature
        coefficient

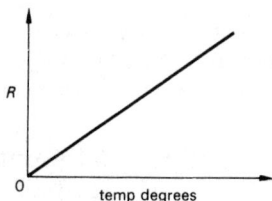

104. In the expression $R_t = R_o(1 + \alpha t)$, R_t is the resistance at temperature t, α is the temperature coefficient, R_o is is
   (a) resistance at $100^{\circ}C$
   (b) final resistance value
   (c) resistance at room temperature
   (d) resistance at $0^{\circ}C$.

105. In the expression $R_t = R_o(1 + \alpha t)$ it is possible to express α as
   (a) $\alpha = \dfrac{R_t}{R_o} - t$
   (c) $\alpha = \dfrac{R_t + R_o}{t}$
   (b) $\alpha = \dfrac{R_t - R_o}{t}$
   (d) $\alpha = \dfrac{R_t - R_o}{R_o t}$.

The construction and principles of operation of moving coil and moving iron type of instruments

106. The restoring force in a moving iron meter is
   (a) gravity   (c) eddy currents
   (b) hairsprings   (d) back e.m.f.

107. The damping force used in a moving coil meter is
   (a) a small phosphor bronze spring on the needle
   (b) a small phosphor bronze spring on the coil
   (c) eddy currents in the coil former
   (d) friction damper on the former mounting.

108. A moving iron meter can be used to measure
   (a) d.c. only
   (b) d.c. and low frequency a.c.,
   (c) a.c. only
   (d) d.c. and a.c. at all frequencies.

109. The forces acting on a meter when measuring current are
   (a) responding, deflection and damping
   (b) deflection, controlling and damping
   (c) reaction, controlling and responding
   (d) centrifugal, gravity and deflection.

110. An ammeter is inserted into the circuit shown. To measure the current accurately the ammeter resistance (R_m) should be
(a) $R_1 + R_2$
(b) $\dfrac{R_1 R_2}{R_1 + R_2}$
(c) very small
(d) infinite.

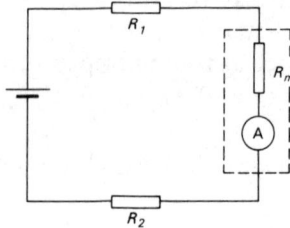

111. A voltmeter of resistance 20 Ω is connected across AB. The voltmeter indicates
(a) 30 V
(b) 40 V
(c) 20 V
(d) 2 V.

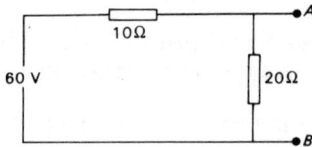

112. The deflection of a moving coil meter is proportional to the
(a) square of the current in the movement and the field strength of the permanent magnet
(b) resistance of the movement and the applied voltage
(c) square of the magnetic flux of the field magnet and the current
(d) current in the coil.

113. A moving coil meter having a full scale deflection of 5 mA is required to operate in a circuit where the maximum value of the current is 25 mA. In this case
(a) a resistor should be placed in series with the meter
(b) the movement should be mechanically damped
(c) a resistor should be placed in parallel with the meter
(d) a capacitor should be placed in series with the meter.

114. A basic meter movement has a full scale deflection of 10 mA and a resistance of 10 Ω. What value of multiplier resistor would be used to convert it to a voltmeter with a full scale deflection of 1 volt?
  (a) 100 Ω
  (b) 90 Ω
  (c) 10 Ω
  (d) 9 Ω.

115. A basic meter movement has a full scale deflection of 1 mA and a resistance of 1 Ω. What value of shunt resistor is required to convert it to an ammeter of full scale deflection 100 mA?
  (a) 10 Ω
  (b) 1 Ω
  (c) 0.01 Ω
  (d) 0.1 Ω.

116. A meter has a full scale deflection of 0.5 mA and a resistance of 10 Ω. To change it to a meter which reads 0 to 10 V the resistor needed is
  (a) 20 000 Ω in series
  (b) 19 990 Ω in series
  (c) 20 000 Ω in parallel
  (d) 19 990 Ω in parallel.

117. The resistance of a moving coil meter with its multiplier is 100 Ω. It requires 5 mA to produce full scale deflection. The meter with its multiplier is most suitable for use as
  (a) 0—5 mA ammeter
  (b) 0—500 V voltmeter
  (c) 0—5 V voltmeter
  (d) 0—500 mA ammeter.

118. A moving coil meter of resistance 100 Ω requires 10 mA to deflect full scale. What value of resistance will enable the meter to indicate full scale when 10 V is applied?
  (a) 900 Ω   (c) 1000 Ω
  (b) 100 Ω   (d) 90 Ω.

119. A 2 Ω resistor connected across a moving coil meter causes the current range to be doubled. What is the resistance of the meter?

    (a) 5 Ω
    (b) 10 Ω
    (c) 2 Ω
    (d) 0.5 Ω.

120. An ammeter of resistance 10 Ω requires 50 mA for full scale deflection. What value of shunt is required so that the meter has a full scale deflection of 550 mA?

    (a) $\dfrac{10}{11}$ Ω
    (b) $\dfrac{11}{10}$ Ω
    (c) 1 Ω
    (d) 10 Ω.

CHAPTER TWO

Electrical Principles – Level II

CHAPTER OBJECTIVES

After studying this chapter you should be able to:

* solve problems involving series-parallel resistive circuits (121–125);
* use Kirchhoff's laws to solve restrictive network problems involving no more than two simultaneous equations (126–135);
* explain the laws relating to magnetic fields and apply them to series magnetic circuits (136–150);
* relate the laws of electromagnetic induction to electrical circuits, and to the principles of generators and transformers (151-163);
* explain the property of capacitance of a parallel plate capacitor (164–172);
* determine the equivalent capacitance of a number of capacitors connected in series and parallel (173–177);
* derive an expression for the energy stored in a charged capacitor (178–181);
* carry out calculations involving the fundamentals of alternating current theory (182–200);
* understand the behaviour of simple series a.c. circuits (201–225);
* describe the construction, principles of action and use of measuring instruments of the scale and pointer type (226–233);
* modify a moving coil instrument to measure alternating quantities (234–236);
* explain the principle of the Wheatstone bridge and a.c. potentiometer (237–240).

Problems involving series-parallel resistive circuits

121. The potential difference across the unknown resistor R
is
(a) 180 V
(b) 18 V
(c) 20 V
(d) 2 V.

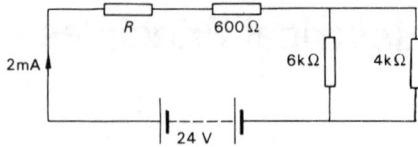

122. If the current in the 20 Ω resistor is 1 A the applied
voltage E is
(a) 60 V   (c) 65 V
(b) 95 V   (d) 165 V.

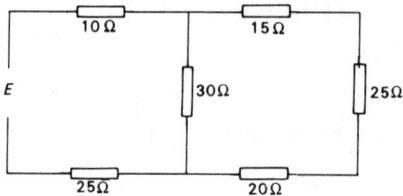

123. In the figure the resistors are connected so that
(a) R_1 and R_3 are in parallel with each other, but in series
with R_2 and R_4
(b) R_1 and R_2 are in parallel with each other, but in series
with R_3 and R_4
(c) R_1, R_2 and R_4 are in parallel with each other, but
in series with R_2
(d) There are three parallel branches containing R_1 in
the first, R_3 in the second and R_2 and R_4 in series
in the third.

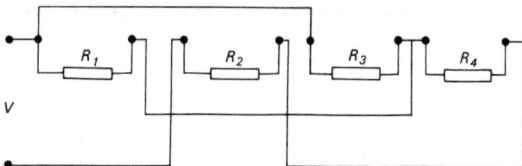

124. If the potential difference across the 20 Ω resistor is 32 V the potential difference across the 40 Ω resistor will be
    (a) 32 V
    (b) 16 V
    (c) 48 V
    (d) 64 V.

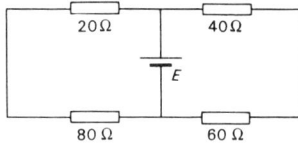

125. In the figure the value of the supply current I is
    (a) 1 A
    (b) 2 A
    (c) 3 A
    (d) 4 A.

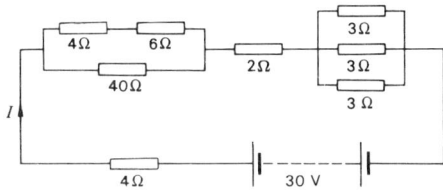

The use of Kirchhoff's laws to solve resistive network problems involving no more than two simultaneous equations

126. Kirchhoff's voltage law states that
    (a) the voltage drop across a resistor is equal to the current multiplied by the resistance
    (b) the total voltage drop in a series circuit cannot be zero
    (c) the algebraic sum of the currents meeting at a junction is zero
    (d) the algebraic sum of the e.m.f.s. and the potential differences in a series circuit must be zero.

127. Kirchhoff's current law states that
    (a) the algebraic sum of the currents entering a junction is zero
    (b) the algebraic sum of the currents in a series circuit is zero
    (c) the algebraic sum of the currents in a parallel circuit is zero
    (d) the sum of the e.m.f.s. and the current in a closed circuit are equal.

128. In the circuit shown, which of the following equations is correct?

    (a) $25 = 2i_1 + 5(i_1 + i_2)$

    (b) $10 = 5i_1 + 2(i_1 + i_2)$

    (c) $15 = 2i_1 + 5(i_1 + i_2)$

    (d) $i_2 = 7(i_1 + i_2)$.

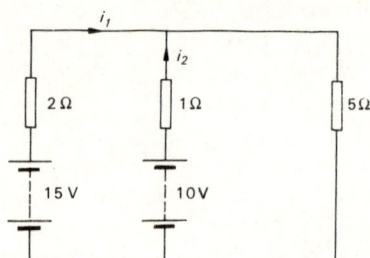

129. Five secondary cells each of e.m.f. 2 V are connected in series. The internal resistance of each cell is 0.02 Ω. If the value of the load resistance is 1.9 Ω the total current is

    (a) 5.2 A   (c) 1 A

    (b) 5 A   (d) 1.05 A.

130. At the junction shown the value of I is

    (a) $I_1 + I_3 - I_2$

    (b) $I_2 - I_3 - I_1$

    (c) $I_1 + I_2 + I_3$

    (d) $I_1 - I_3 + I_2$.

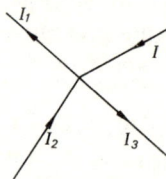

131. In the diagram R and $2R$ are resistors in parallel. The current I_1 is

    (a) $\frac{1}{2}I$

    (b) $\frac{1}{3}I$

    (c) $\frac{2}{3}I$

    (d) $2I_2$.

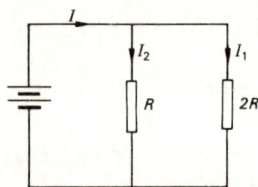

132. In the circuit the potential difference between terminals
     A and B is
     (a) 2.5 V
     (b) 7.5 V
     (c) 27.5 V
     (d) 12.5 V.

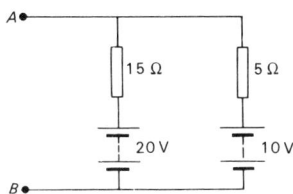

133. The potential of A with respect to earth is
     (a) −3¾ V
     (b) −6 V
     (c) +3 V
     (d) +21¾ V.

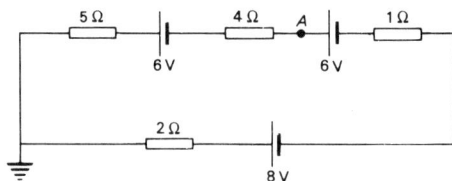

134. If in the circuit shown $V_1 = 2V_2$ the resistance R is
     (a) 7.5 Ω
     (b) 5.0 Ω
     (c) 15 Ω
     (d) 20 Ω.

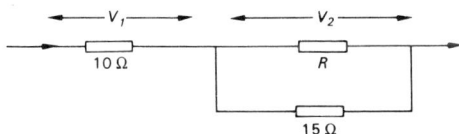

135. In the system of conductors shown, the currents i_1 and
     i_2 are
     (a) $i_1 = 4$ A, $i_2 = -8$ A
     (b) $i_1 = 16$ A, $i_2 = 20$ A
     (c) $i_1 = -4$ A, $i_2 = 0$
     (d) $i_1 = 4$ A, $i_2 = 12$ A.

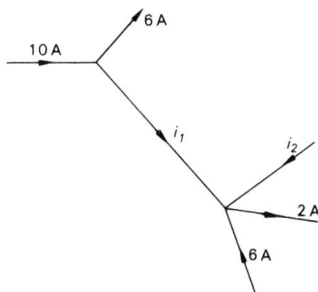

The laws relating to magnetic fields and applying them to series magnetic circuits

136. A straight wire carrying a direct current is at 90° to the plane of the paper, the current flow being into the paper. The magnetic lines of force are represented as
(a) radial lines away from the wire
(b) concentric circles having clockwise direction
(c) concentric circles having anticlockwise direction
(d) radial lines towards the wire.

137. A current-carrying conductor is placed between the pole pieces of a magnet as shown. The conductor will tend to move
(a) towards the N pole
(b) towards the S pole
(c) to the right
(d) to the left.

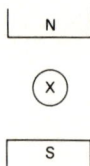

138. The magnitude of the magnetic flux passing through 1 square metre is called
(a) flux density
(b) magnetomotive force
(c) magnetic circuit
(d) residual magnetism.

139. A force of 10 N is experienced by a conductor 1 metre long as it moves through a magnetic field of density 0.5 T. The current flowing in the conductor is
(a) 0.05 A
(b) 10 A
(c) 0.5 A
(d) 20 A.

140. When an alternating current is passed through the coil of the electromagnet the piece of soft iron will be
    (a) attracted towards the electromagnet
    (b) repelled away from the electromagnet
    (c) remain unaffected by the electromagnet
    (d) alternately attracted and repelled depending on the direction of the current.

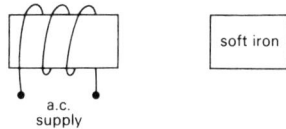

141. When nearing magnetic saturation the relative permeability of iron
    (a) decreases
    (b) increases then falls rapidly
    (c) remains the same
    (d) increases.

142. In a magnetic circuit S represents reluctance; Φ, magnetic flux; m.m.f. magnetomotive force; and B, flux density; Φ can be expressed as
    (a) $\Phi = \dfrac{\text{m.m.f.}}{S}$
    (b) $\Phi = S \times \text{m.m.f.}$
    (c) $\Phi = \dfrac{R}{\text{m.m.f.}}$
    (d) $\Phi = B \times S$.

143. A current of 10 A flows through a coil of 200 turns uniformly wound on an iron ring of mean circumference 50 cm. The magnetic field strength H is
    (a) 10 A/m
    (b) 4000 A/m
    (c) 20 000 At/m
    (d) 200 turns.

144. The reluctance of S of a magnetic circuit of length ℓ, cross-sectional area a, and permeability μ, can be expressed as:

(a) $S = \dfrac{\mu\ell}{a}$ (c) $S = \dfrac{a}{\mu\ell}$

(b) $S = \dfrac{\ell}{\mu a}$ (d) $S = \dfrac{a\ell}{\mu}$.

145. A coil of 200 turns produces a flux of 0.2 mWb when a current of 2 A flows in the coil. The reluctance of the magnetic circuit is

(a) 2×10^6 A/Wb   (c) 1 A/Wb

(b) 2 A/Wb   (d) 2 A.

146. A simple magnetic circuit lies entirely in iron. An air gap is introduced into the circuit. It is found that the reluctance

(a) increases

(b) decreases

(c) remains unaltered

(d) increases by a factor of two.

147. Two dissimilar metals form a composite magnetic circuit as shown. The total relutance of the circuit is given by

(a) $S_T = \dfrac{\ell_1}{\mu_1 a_1} - \dfrac{\ell_2}{\mu_2 a_2}$

(b) $S_T = \dfrac{\ell_2}{\mu_2 a_2} - \dfrac{\ell_1}{\mu_1 a_1}$

(c) $S_T = \dfrac{\ell_1 \, \ell_2}{\mu_1 \mu_2 a_1 a_2}$

(d) $S_T = \dfrac{\ell_1}{\mu_1 a_1} + \dfrac{\ell_2}{\mu_2 a_2}$.

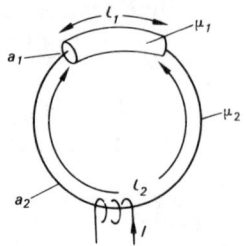

148. Examination of the hysteresis curve in the figure shows that

(a) Y has a higher remanance and coercivity than X

(b) Y has a higher remanance and lower coercivity than X

(c) X has a higher remanance and lower coercivity than Y

(d) X has a lower remanance and higher coercivity than Y.

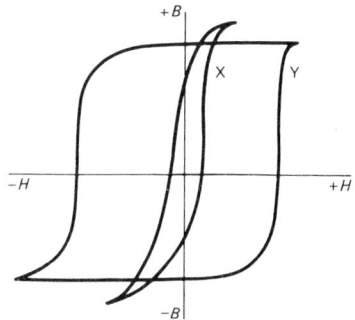

149. The intercept on the B axis of a hysteresis loop for a steel specimen which has been taken to saturation indicates
    (a) the remanance of the specimen
    (b) the coercivity of the specimen
    (c) the work done in taking the specimen through a cycle of magnetism
    (d) the permeability of the specimen.

150. If the area enclosed by a hysteresis curve is large the energy loss will be
    (a) low
    (b) high
    (c) inversely proportional to area
    (d) proportional to area^2.

Relation of the laws of electromagnetic induction to electrical circuits, and to the principles of generators and transformers

151. The e.m.f. induced in a coil due to changing current is:
    (a) proportional to the inductance and the rate of change of current
    (b) inversely proportional to the inductance and the rate of change of current
    (c) proportional to the resistance of the coil and the rate of change of current
    (d) determined by Ohm's law.

152. The self-inductance of a coil is 1 henry when
   (a) a current changing at the rate of 1 A per second induces an e.m.f. of 1 volt in a second coil
   (b) a current of 1 A induces an e.m.f. of 1 volt
   (c) an e.m.f. changing at 1 volt per second induces a current of 1 A
   (d) a current changing at the rate of 1 A per second induces an e.m.f. of 1 volt in that coil.

153. The current flowing in an inductance is increased uniformly from 0 to 2 amps in one-tenth of a second. The induced e.m.f. is 10 V. The current is now increased uniformly from 2 amps to 4 amps in a further period of one-tenth of a second. The induced e.m.f. will be
   (a) 10 V
   (b) 20 V
   (c) 5 V
   (d) 40 V.

154. A coil of fifty turns is found to produce a flux of 2 mWb when the current flowing is 1 amp. Its inductance will be
   (a) 10 mH
   (b) 100 mH
   (c) 25 mH
   (d) 250 mH.

155. If the number of turns in an inductor is doubled the inductance value will be found to approximately
   (a) increase by a factor of four
   (b) decrease by a factor of four
   (c) double
   (d) halve.

156. The energy stored in an inductor can be found from the formulae:
   (a) $E = \dfrac{L}{I^2}$
   (b) $E = \frac{1}{2} LI$
   (c) $E = \frac{1}{2} LI^2$
   (d) $E = \frac{1}{2} L^2 I$.

157. If the current flowing in an inductor is kept constant and the number of turns increased ten times, the energy stored in the magnetic field will
    (a) remain unchanged
    (b) increase 10 times
    (c) increase 50 times
    (d) increase 100 times.

158. Two circuits have a mutual inductance of 1 henry if an e.m.f. of 1 volt is produced in one circuit when the current in the other varies at a rate of
    (a) 1 amp per second
    (b) 1 amp per second squared
    (c) 1 metre per second
    (d) 1 milliamp per second.

159. If in a transformer the number of primary turns is doubled and the number of secondary turns is halved, the mutual inductance between primary and secondary will
    (a) double
    (b) halve
    (c) remain constant
    (d) increase by a factor of four.

160. Two coils are mutually couped with a coupling factor of 0.5. If each of the coils has a self-inductance of 100 mH the mutual inductance of the two coils will be
    (a) 100 mH
    (b) 50 mH
    (c) 10 000 mH
    (d) 5000 mH.

161. Two coils are wound on an iron ring; if the ring was replaced by one made of polystyrene the mutual inductance of the two coils would
    (a) remain unaffected
    (b) increase slightly
    (c) decrease
    (d) double.

162. A loss free transformer has 1000 turns in its secondary and 100 turns in its primary. If the current in the primary changes at 2 amps per second the current in the secondary will change at
    (a) 20 amps per second
    (b) 2 amps per second
    (c) 0.2 amps per second
    (d) 0.02 amps per second.

163. The Time Constant (T.C.) of a circuit containing a resistor R and inductance L is given by
    (a) T.C. = LR
    (b) T.C. = $\dfrac{R}{L}$
    (c) T.C. = $\dfrac{L}{R}$
    (d) T.C. = $\dfrac{L^2}{R}$.

The property of capacitance of a parallel plate capacitor

164. If the distance between the plates of a parallel plate capacitor is halved and the area of the plates is doubled then the capacitance would be
    (a) doubled
    (b) increased eight times
    (c) unaltered
    (d) increased four times.

165. An 80 pF parallel plate capacitor has the area of its plates halved and the distance between them increased four times. The resulting capacitance will be
    (a) 10 pF
    (b) 40 pF
    (c) 160 pF
    (d) 1000 pF.

166. A 30 μF capacitor is charged from a 400 V supply. The charge stored by the capacitor is
    (a) 13.3 C   (c) 1.3 MC
    (b) 12 000 C   (d) 12 mC.

167. Electrolytic capacitors are employed where one of the following is required
    (a) high capacitance and small size
    (b) low inductance
    (c) high breakdown voltage
    (d) low leakage current.

168. If large values of capacitance are required the following capacitor is usually used
    (a) paper
    (b) mica
    (c) air
    (d) electrolytic.

169. Which type of capacitor is unsuitable for use in circuits in which the potential difference across the capacitor is likely to reverse in polarity?
    (a) preset
    (b) electrolytic
    (c) mica
    (d) paper.

170. A parallel plate capacitor has a capacitance of 1 μF when measured in a vacuum. If a sheet of mica is then placed between the plates the capacitance will
    (a) increase
    (b) decrease
    (c) remain unaltered
    (d) decrease by a factor
        of two.

171. The plates of a parallel plate capacitor are separated by a distance of 10 mm. If the potential difference across the plates is 20 V the electrical stress (E) will be
    (a) 2 mV/m
    (b) 0.5 mV/m
    (c) 500 V/m
    (d) 2000 V/m.

172. The capacitance C of a parallel plate capacitor can be expressed in terms of area of plate A, distance between the plates d, absolute permittivity ϵ, as follows

(a) $C = \dfrac{A}{\epsilon d}$

(b) $C = \dfrac{\epsilon A}{d}$

(c) $C = \dfrac{\epsilon d}{A}$

(d) $C = \dfrac{Ad}{\epsilon}$.

To determine the equivalent capacitance of a number of capacitors connected in series and parallel

173. When two capacitors are connected in series
   (a) the charge stored on each is proportional to its capacitance
   (b) the charge stored on each is inversely proportional to its capacitance
   (c) the charge stored on each is the same
   (d) the charge stored on each is inversely proportional to the voltage across each.

174. In the diagram shown the effective capacitance between A and B is
   (a) $10\ \mu F$
   (b) $1\ \mu F$
   (c) $4\ \mu F$
   (d) $1.6\ \mu F$.

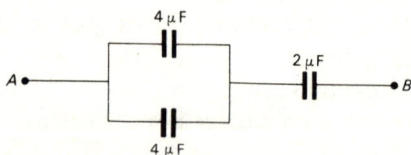

175. Five identical capacitors of 20 pF in value are connected as shown. The total capacitance between A and B is approximately
   (a) 23 pF
   (b) 70 pF
   (c) 17 pF
   (d) 100 pF.

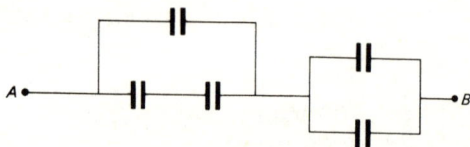

176. The two capacitors shown are charged to 10 V. On closing the switch S

   (a) conventional current will flow from A to B

   (b) conventional current will flow from B to A

   (c) the total charge stored will decrease

   (d) no current flow will occur.

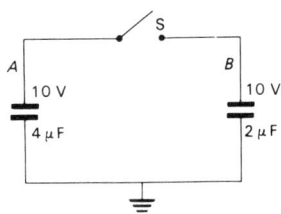

177. If the safe working voltages of the 4 μF and 6 μF capacitors shown are 12 V and 20 V respectively, the maximum allowable potential difference between A and B is

   (a) 18 V
   (b) 32 V
   (c) 12 V
   (d) 20 V.

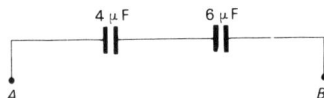

Derivation of an expression for the energy stored in a charged capacitor

178. If the voltage across a capacitor is doubled the energy stored will

   (a) double
   (b) halve
   (c) increase four times
   (d) decrease by a factor of four.

179. A 2 μF capacitor is charged to 1000 V, the energy stored will be

   (a) 2 joules
   (b) 1 joule
   (c) 1000 joules
   (d) 500 joules.

180. If the total energy stored by the two 4 μF capacitors shown is 2500 J, what is the value of V?
    (a) 50 V
    (b) 500 V
    (c) 625 V
    (d) 5000 V.

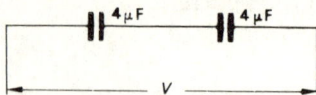

181. The energy stored in the 10 μF capacitor in the circuit shown is
    (a) 50 J
    (b) 100 J
    (c) 100 mJ
    (d) 50 mJ.

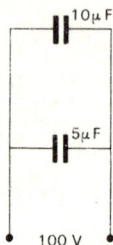

Calculations involving the fundamentals of alternating current theory

182. The period of a sine wave is 1 ms. The frequency is
    (a) 100 Hz
    (b) 500 Hz
    (c) 1 kHz
    (d) 2 kHz.

183. A sine wave passes through zero once every millisecond. The frequency of the waveform is
    (a) 200 Hz
    (b) 500 Hz
    (c) 1 kHz
    (d) 2 kHz.

184. The frequency of the output from an alternator which has eight poles and is rotating at 750 revolutions per minute is
    (a) 60 Hz
    (b) 50 Hz
    (c) 94 Hz
    (d) 2000 Hz.

185. A sinusoidal alternating current is measured with an a.c. ammeter and reads 100 A. Its peak value is
    (a) 100 A
    (b) 141.4 A
    (c) 63.7 A
    (d) 70.7 A

186. The r.m.s. values of a sinusoidal alternating supply are 1000 V and 6 A. The peak values are
    (a) 1414 V and 8.484 A
    (b) 707 V and 4.242 A
    (c) 707 V and 8.484 A
    (d) 1414 V and 4.242 A.

187. The peak values of an alternating supply are 141.4 V and 10 A. The r.m.s. values are
    (a) 70.7 V and 7.07 A
    (b) 200 V and 14.14 A
    (c) 50 V and 3.04 A
    (d) 100 V and 7.07 A.

188. An alternating sinusoidal voltage is displayed on an oscilloscope. If the peak to peak voltage measured is 50 V the r.m.s. value of voltage will be
    (a) 17.7 V
    (b) 25 V
    (c) 35.4 V
    (d) 70.7 V.

189. The resistance of the element of a 2 kilowatt electric fire operated from 250 V, 50 Hz supply mains is
    (a) 177.45 Ω
    (b) 62.5 Ω
    (c) 31.25 Ω
    (d) 17.74 Ω.

190. The vector diagram shown represents two alternating voltages
    (a) V_1 lags V_2 by 135°
    (b) V_1 and V_2 are in antiphase
    (c) V_2 leads V_1 by 45°
    (d) V_2 lags V_1 by 135°.

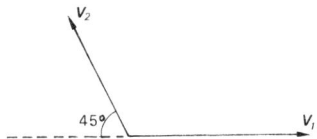

191. The vector diagram shown represents an alternating voltage and current where
  (a) **I** lags **V** by 15°
  (b) **I** leads **V** by 15°
  (c) **V** lags **I** by 75°
  (d) **V** leads **I** by 75°.

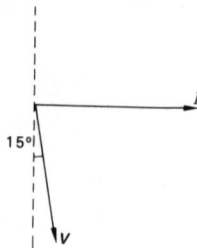

192. The r.m.s. value of the resultant of two sine waves of equal ampitude but differing in phase by 90° is
  (a) zero
  (b) twice the r.m.s. value of one of the waves
  (c) half the r.m.s. value of one of the waves
  (d) 1.414 times the r.m.s. value of either of the waves.

193. In an a.c. circuit the current leads the voltage by 90°. This circuit is behaving as a pure
  (a) resistance
  (b) capacitance
  (c) inductance
  (d) impedance.

194. The relationship between voltage and current in a purely capacitive circuit is
  (a) current and voltage are in phase
  (b) current leads voltage by 90°
  (c) voltage leads current by 90°
  (d) voltage and current are in antiphase.

195. If the frequency of the voltage applied to an inductor is increased then
  (a) its reactance will fall
  (b) its reactance will rise
  (c) the current through the inductor will rise
  (d) the current through the coil will remain unchanged.

196. The reactance of a 100 microhenry inductor at 250 kHz
     is approximately
     (a) 25 Ω   (c) 160 Ω
     (b) 0.25 Ω   (d) 0.00016 Ω.

197. A capacitor of 0.003 μF has a certain reactance at a
     frequency of 400 Hz. The value of capacitor which will
     give approximately the same reactance at a frequency of
     1600 Hz is
     (a) 0.012 μF
     (b) 0.00075 μF
     (c) 0.00125 μF
     (d) 0.0067 μF.

198. An alternating supply of constant voltage and variable
     frequency is applied to a purely resistive circuit. The
     frequency is varied from 50 Hz to 4 kHz. The r.m.s.
     value of the current will
     (a) increase
     (b) decrease
     (c) fluctuate
     (d) not change.

199. The graph showing the variation of inductive reactance
     with frequency is a
     (a) sine curve
     (b) cosine curve
     (c) rectangular hyperbola
     (d) straight line.

200. If the frequency of the a.c. voltage applied to a
     capacitor is increased the reactance of the capacitor
     will
     (a) increase
     (b) decrease
     (c) remain constant
     (d) increase as the square of the frequency.

The behaviour of simple series a.c. circuits

201. If the frequency of the voltage applied between A and B
     is adjusted until the reactance of the capacitor equals

the resistance in the circuit the impedance between A and B will be

(a) 0 Ω

(b) 1414 Ω

(c) 2000 Ω

(d) 1 MΩ.

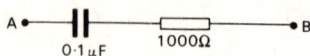

A •—||—[1000Ω]—• B
0·1 μF   1000Ω

202. The voltage across the inductor in the circuit shown is

(a) 25 mV leading voltage across the resistor by 90°

(b) 25 V in phase with the voltage across resistor

(c) 50 V in phase with the voltage across resistor

(d) 50 V leading the voltage across the resistor by 90°.

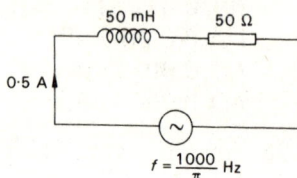

50 mH   50 Ω
0·5 A
$f = \frac{1000}{\pi}$ Hz

203. A series circuit containing capacitance and resistance is connected across an alternating supply. The voltage across the capacitor

(a) lags the applied voltage by 90°

(b) is in phase with the voltage across the resistor;

(c) leads the applied voltage by 90°

(d) lags the current by 90°.

204. In the vector diagram shown, where X_L = 12 Ω and R = 5 Ω, the resultant impedance and phase angle would be

(a) 17 Ω and $\tan^{-1}$ 0.416

(b) 19 Ω and $\cos^{-1}$ 0.385

(c) 13 Ω and $\sin^{-1}$ 0.385

(d) 17 Ω and $\tan^{-1}$ 2.4.

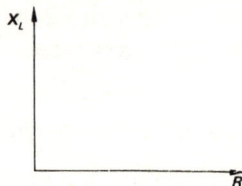

X_L
R

205. In the circuit shown the component X has a reactance
   of 10 kΩ and I lags V by 45° X is therefore
   - (a) an inductor and R is
     10 kΩ
   - (b) an inductor and R is
     20 kΩ
   - (c) a capacitor and R is
     10 kΩ
   - (d) a capacitor and R is
     20 kΩ.

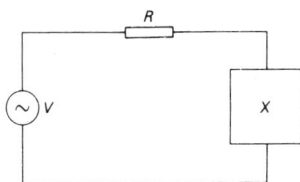

206. In the circuit shown the component X has a reactance
   of 5 kΩ and I leads V by 45°. X is
   - (a) an inductor and R is
     5 kΩ
   - (b) an inductor and R is
     10 kΩ
   - (c) a capacitor and R is
     5 kΩ
   - (d) a capacitor and R is
     10 kΩ.

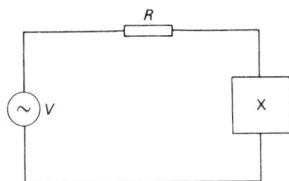

207. In the circuit shown, if the frequency of the generator
   were decreased, the
   - (a) CR time would be
     decreased
   - (b) phase difference
     between V_C and V_R
     would be increased
   - (c) impedance of the
     circuit would be
     increased
   - (d) impedance of the
     circuit would be
     decreased.

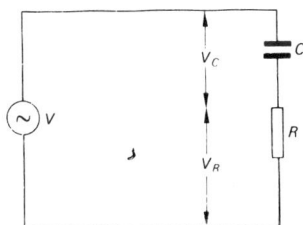

208. In the circuit shown, the amplitude of the applied voltage V is kept constant, but the frequency is decreased. The current is
   (a) directly proportional to the frequency
   (b) inversely proportional to the frequency
   (c) proportional to the product of R and frequency
   (d) inversely proportional to the impedance.

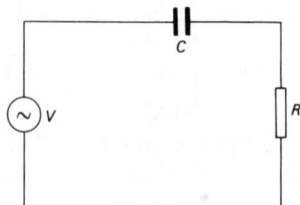

209. If the impedance of the circuit shown is Z then
   (a) $Z = X_L + X_C$
   (b) $Z = X_L - X_C$
   (c) $Z = \sqrt{X_L{}^2 + X_C{}^2}$
   (d) $Z = \sqrt{X_L{}^2 - X_C{}^2}$.

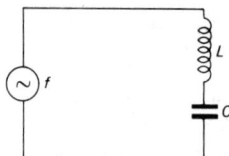

210. The impedance of the circuit shown is
   (a) 7 kΩ
   (b) 12 kΩ
   (c) 5 kΩ
   (d) 1.7 kΩ.

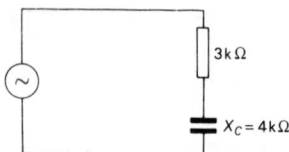

211. In the circuit shown the potential difference across the resistor is 50 V. The potential difference across the capacitor is
   (a) 50 V
   (b) 86.6 V
   (c) 150 V
   (d) 70.7 V.

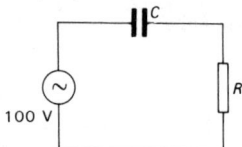

212. In a series LR circuit, X_L = 12 kΩ, R = 5 kΩ and I = 10 mA. The supply voltage is
    (a) 170 V   (c) 70 V
    (b) 130 V   (d) 169 V.

213. The power factor of a pure inductance is
    (a) 1   (c) zero
    (b) 0.707   (d) 0.5.

214. A 16 mH inductor has a resistance of 30 Ω. When connected to a 400 Hz voltage supply the power factor will be
    (a) 0.6   (c) 0.75
    (b) 0.8   (d) 1.3.

215. 100 V is applied to a capacitor which has a resistance of 50 Ω. The true power developed in the capacitor will be
    (a) 200 W   (c) 2 kW
    (b) zero   (d) 100 W.

216. The power factor of this circuit is approximately
    (a) 1.8
    (b) 0.8
    (c) zero
    (d) 0.6.

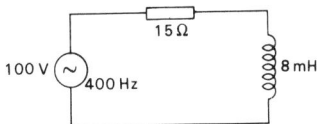

217. The power factor of a circuit may be defined as
    (a) true power ÷ apparent power
    (b) apparent power ÷ true power
    (c) (supply current)^2 $\times$ resistance
    (d) (supply voltage)^2 $\times$ resistance.

218. In a series LCR circuit, R = 50 kΩ, X_C = 50 kΩ, X_L = 50 kΩ and the applied voltage is 300 V. The current in the circuit will be
    (a) 2 mA
    (b) 3 mA
    (c) 5 mA
    (d) 6 mA.

219. In an a.c. series circuit, the current I
   (a) is in phase with E_R
   (b) is antiphase with E_R
   (c) lags E_C by 90°
   (d) leads E_L by 90°.

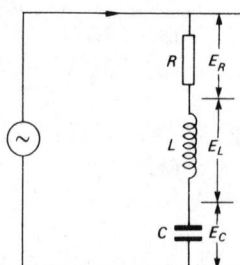

220. In the circuit shown, the amplitude of the applied voltage remains constant and the frequency is increased from zero. V_C will
   (a) increase as frequency increases
   (b) decrease as frequency rises
   (c) decrease to a minimum and then increase
   (d) increase to a maximum and then decrease.

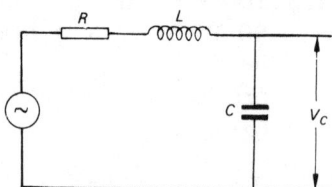

221. The impedance Z of the circuit shown is given by
   (a) $\sqrt{R^2 + X_L^2 - X_C^2}$
   (b) $R + X_L + X_C$
   (c) $R + X_L - X_C$
   (d) $\sqrt{R^2 + (X_L - X_C)^2}$.

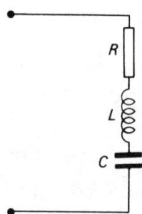

222. In a series LCR circuit the capacitive reactance is 50 ohms greater than the inductive reactance, and the circuit has a resistance of 50 ohms. With respect to the applied voltage the current will
   (a) lead by less than 90°
   (b) lag by less than 90°
   (c) lead by 90°
   (d) be in phase.

223. In the circuit shown the impedance is
    (a) 2500 Ω and the current leads the applied voltage
    (b) 300 Ω and the current lags the applied voltage
    (c) 700 Ω and the current leads the applied voltage
    (d) 500 Ω and the current leads the applied voltage.

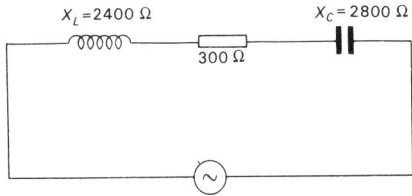

$X_L = 2400\ \Omega$ 300 Ω $X_C = 2800\ \Omega$

224. The impedance of the circuit shown is
    (a) 17 Ω
    (b) 37 Ω
    (c) 13 Ω
    (d) 12 Ω.

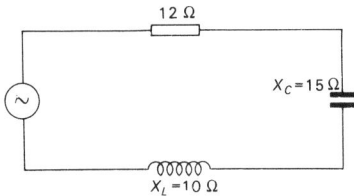

12 Ω $X_C = 15\ \Omega$ $X_L = 10\ \Omega$

225. If the potential differences across the components in the circuit are as shown the generator voltage V must be
    (a) 50 V
    (b) 67 V
    (c) 70 V
    (d) 110 V.

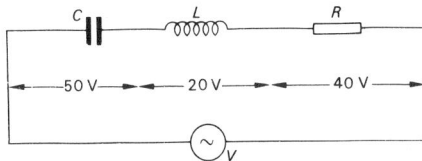

C   L   R
—50 V—   — 20 V—   — 40 V —
V

The construction, principle of action and use of measuring instruments of the scale and pointer type

226. In a moving coil meter the controlling force is provided by
    (a) the current being measured passing through the coil
    (b) eddy currents in the aluminium former
    (c) springs which also conduct current through the coil
    (d) eddy currents in the soft iron core.

227. Damping in a moving coil meter is obtained by the
     (a) action of the hairsprings
     (b) interaction of the coil current and the magnetic flux
     (c) eddy currents in the magnet
     (d) eddy currents in the coil former.

228. When ammeters or voltmeters are introduced into a circuit they should have very little effect on that circuit. It follows, therefore, that
     (a) an ammeter has a large resistance so that very little current is taken
     (b) a voltmeter has a large resistance so that very little current is taken
     (c) a voltmeter has a small resistance so that there is a very small voltage drop across it
     (d) the resistance of a voltmeter is unimportant as the voltage drop across it is always the same as the applied voltage.

229. The current drawn by a voltmeter from a circuit under test should be
     (a) the whole of the circuit current if a true reading is required
     (b) as near to zero as possible
     (c) greater than the full scale deflection current of the meter
     (d) less than the actual p.d. being measured.

230. A moving coil meter is rated at 1000 Ω/V. The resistance of the meter is 19 Ω. If a 1 Ω resistor is connected in parallel with the meter the maximum value that the meter can now indicate is
     (a) 20 mA
     (b) 1 mA
     (c) 19 mA
     (d) 20 mV.

231. A meter of resistance 30 Ω and requiring 1 A to produce full scale deflection is to be adapted to indicate

currents from 0 to 10 A. The value of the shunt required is
(a) 3 Ω
(b) 9 Ω
(c) 20 Ω
(d) 0.5 Ω.

232. A meter has a resistance of 5 Ω and a scale which reads 0 to 1 A. To convert this meter to read 0—100 V the value of multiplier resistance needed is
(a) 0.005 Ω
(b) 9.995 Ω
(c) 95 Ω
(d) 100 Ω.

233. The higher the voltage range in use in an AVO meter, the greater is the multiplier resistance in circuit. On the highest possible voltage range the AVO will
(a) increase the chances of cutout operation
(b) cause least disturbance to circuit conditions
(c) take longer to indicate the p.d. being measured
(d) cause most disturbance of circuit conditions.

234. If a moving coil meter is connected to a 50 Hz a.c. supply it does not give a reading unless a rectifier is included in the circuit because
(a) alternating current cannot flow in a d.c. circuit
(b) the inductance of the coil is too high
(c) alternating current in the coil cannot produce a magnetic field
(d) the pointer cannot move quickly enough to follow changes in the current direction.

235. A rectified moving coil ammeter when connected to an a.c. supply will measure
(a) peak current
(b) r.m.s. current
(c) average current
(d) power factor.

236. To measure the alternating current in the circuit shown the ammeter will be a
   (a) rectified moving coil meter
   (b) ballistic galvanometer
   (c) electrostatic meter
   (d) moving coil ammeter.

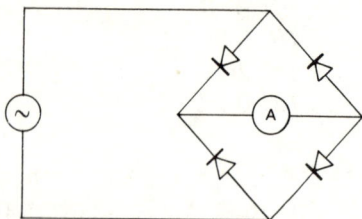

The principle of the Wheatstone bridge and a.c. potentiometer

237. In the circuit shown the variable resistor has to be set at 500 Ω for the ammeter to read zero volts. The value of R must be
   (a) 600 Ω
   (b) 2.4 Ω
   (c) 3 Ω
   (d) 48 Ω.

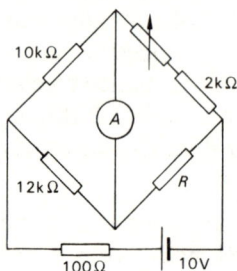

238. In the circuit shown, if the potential difference across the 10 Ω resistor is 16 V the potential difference across the 20 Ω resistor must be
   (a) 8 volts
   (b) 16 volts
   (c) 24 volts
   (d) 32 volts.

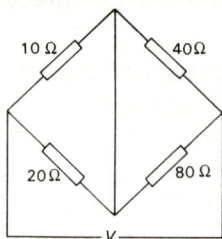

239. Which of the following equations applies when the Wheatstone bridge circuit shown is balanced?

(a) $R_1 R_2 = R_3 R_4$

(b) $R_1 + R_3 = R_2 + R_4$

(c) $R_1 + R_2 = R_3 + R_4$

(d) $R_1 R_4 = R_2 R_3$.

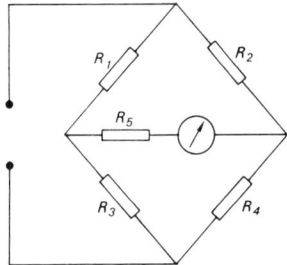

240. The ammeter in the circuit shown will indicate

(a) 0 A

(b) 1¾ ?

(c) 1¼ A

(d) 5 A.

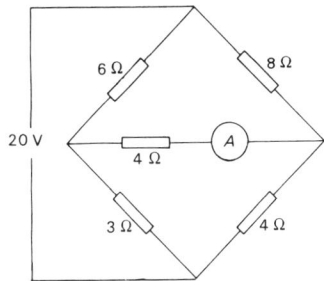

CHAPTER THREE

Electrical Principles – Level III

CHAPTER OBJECTIVES

After studying this chapter you should be able to:

* apply electromagnetic theory to the exponential growth and decay in an inductive-resistive circuit (241–248);
* apply electrostatic theory to the exponential growth and decay in a capacitive-resistive circuit (249–256);
* appreciate the use of "j" notation, in the solution of a.c. series "LCR" circuit problems (257–266);
* understand the significance of series LRC circuits at resonance (267–277);
* appreciate the use of phasor diagrams in the analysis of parallel R, L and C circuits (278–285);
* understand the significance of the parallel resonant condition of an R and L series circuit in parallel with C (286–297);
* understand the basic construction and principles of action of a single phase transformer (298–309);
* understand the construction and actions of a.c. motors and generators (310–320);
* understand the basic construction and action of d.c. motors and generators (321–338);
* understand the basic principles of three-phase systems and solve problems involving balanced systems only (339–348);
* understand the significance of hysteresis in magnetic circuits (349–353);
* understand the concept of energy stored in an electrical field between charged parallel plates (354–357);
* demonstrate a basic knowledge of the construction and principles of the wattmeter (358–360).

The application of electromagnetic theory to the exponential growth and decay in an inductive resistive circuit

241. The "time constant" of a LR circuit connected to a d.c. supply is

  (a) the time required for the current to reach its maximum value

54

(b) the time required for the voltage to reach its maximum value

(c) the time taken for the current to reach 0.632 of its maximum value

(d) the time interval after closing the circuit and before the current actually flows.

242. The time constant of a circuit containing an inductor of 100 mH and a resistor of 5 Ω is

(a) 20 milliseconds

(b) 100 milliseconds

(c) 500 milliseconds

(d) 5 seconds.

243. In the circuit shown, when the switch is closed, the current flowing will rise to 2 A in approximately

(a) 250 seconds

(b) 50 seconds

(c) 2.5 seconds

(d) 0.5 seconds.

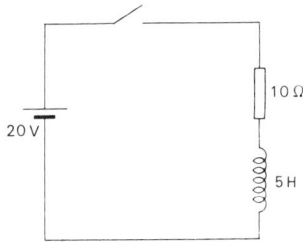

20 V   10 Ω   5 H

244. If the time constant of the circuit shown is 0.04 seconds the value of the inductance is

(a) 4 H

(b) 25 mH

(c) 0.4 H

(d) 0.25 H.

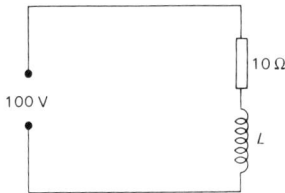

100 V   10 Ω   L

245. The initial rise of current in a *LR* circuit connected to d.c. supply of *V* volts is given by

(a) $\dfrac{V}{L}$

(c) $\dfrac{V}{R}$

(b) $\dfrac{V}{LR}$

(d) $\dfrac{Rt}{L}$.

246. The graph shown represents the rise of current in an LR circuit connected to a d.c. voltage of V volts. The formulae representing the shape of the curve is

(a) $i = I \ (\exp^{-\frac{t}{RL}})$

(b) $i = V \ (\exp^{-\frac{Rt}{L}})$

(c) $i = I \ (\exp^{-\frac{Rt}{L}})$

(d) $i = i \ (\exp^{-\frac{Lt}{R}})$.

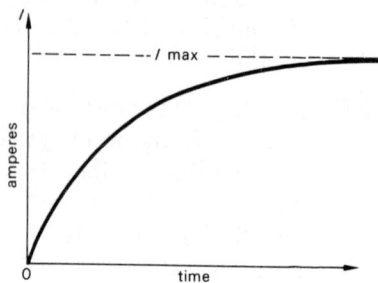

247. At the instant of closing the switch S the current in the circuit will be
   (a) at a minimum value
   (b) at a maximum value
   (c) 20 mA
   (d) 5 mA.

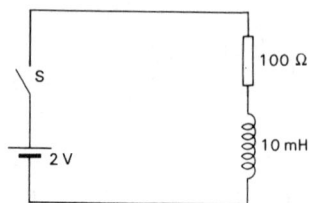

248. Switch S_1 is held in position A for 1 minute and then moved to position B. 20 ms later the current flowing in the inductor will be approximately
   (a) 0.37 A
   (b) 0.63 A
   (c) 0.74 A
   (d) 1.26 A.

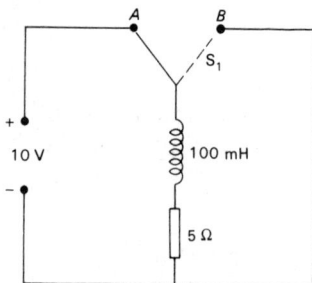

The application of electrostatic theory to the exponential growth and decay in a capacitive resistive circuit

249. When switch S is closed, the time for the capacitor to charge to 63 V is
(a) 5 seconds
(b) 1 second
(c) 5 milliseconds
(d) 2 milliseconds.

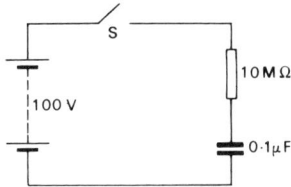

250. In the circuit shown, 100 μs. after the switch is closed the voltage across the capacitor will be
(a) 37 volts
(b) 90 volts
(c) 63 volts
(d) 10 volts.

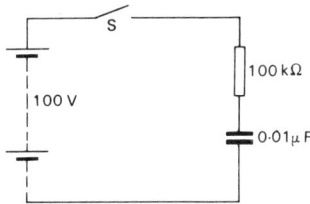

251. In the circuit shown at time $t = 0$, S is closed, thereafter
(a) I grows exponentially
(b) V_C decreases exponentially
(c) I rises instantaneously and then decreases exponentially
(d) charge on C increases linearly.

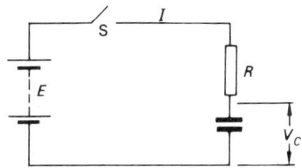

252. On closing the switch S in the circuit shown
(a) the p.d. between A and B rises
(b) the total circuit capacitance falls

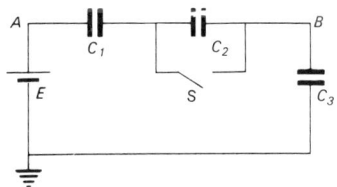

(c) the total energy
stored in the
capacitors falls
(d) the potential at point
B falls.

253. In the circuit shown, on closing the switch:
   (a) p.d. across the 6 μF
       will be 100 V after
       1 second
   (b) p.d. across the 4 μF
       will be 100 V after
       400 milliseconds
   (c) p.d. across the 5 Ω
       resistor will be 36.8 V
       after 200 milliseconds
   (d) p.d. across the 4 μF
       capacitor will be
       63.2 V after 80
       milliseconds.

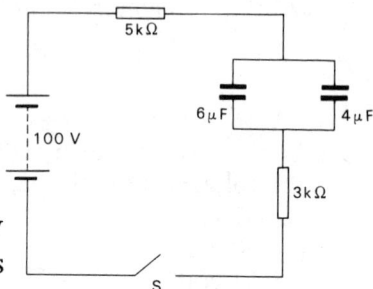

254. On closing the switch in the circuit shown, the milli-
     ammeter deflects rapidly and then returns slowly to
     zero. The unknown circuit element X comprises
   (a) capacitance only
   (b) a series combination
       of inductance and
       resistance
   (c) a series combination
       of capacitance and
       resistance
   (d) a parallel combination
       of resistance and
       capacitance.

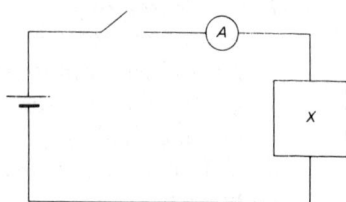

255. The discharge of a capacitor C through a resistor R after
     being charged up to a voltage V is given by

   (a) $v = V \left(\exp^{-\frac{RC}{t}} \right)$

(b) $v = V\left(\exp^{-\frac{t}{RC}}\right)$

(c) $v = -V\left(\exp^{-\frac{t}{RC}}\right)$

(d) $v = V\left(\exp^{-\frac{Ct}{R}}\right)$.

256. A capacitor of $10\,\mu$F is connected in series with a resistor of $10\,\Omega$ and is suddenly connected across a d.c. supply of 1 V. The initial current will be
   (a) 0.1 A
   (b) 0.1 mA
   (c) 0.5 mA
   (d) 10 A.

The use of j notation, in the solution of a.c. series LCR circuit problems

257. The current I flowing in the circuit shown is
   (a) $(33.3 + j\,25)$ A
   (b) $(33.3 - j\,25)$ A
   (c) $(12 - j\,16)$ A
   (d) $(14.29 + j\,0)$ A.

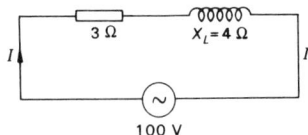

258. The current I flowing in the circuit shown is
   (a) $(1429 \lfloor 0^\circ)$ A
   (b) $(14.29 \lfloor -53.1^\circ)$ A
   (c) $(14.29 \lfloor +53.1^\circ)$ A
   (d) $(20 \lfloor -53.1^\circ)$ A.

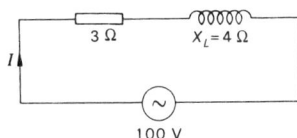

259. A voltage $V = 200 \lfloor 45^\circ$ volts is applied to a circuit. The resultant current is given by $I = 10 \lfloor -10^\circ$. The impedance of the circuit is
   (a) $20 \lfloor 35^\circ\ \Omega$
   (b) $20 \lfloor 55^\circ\ \Omega$
   (c) $0.05 \lfloor -55^\circ\ \Omega$
   (d) $0.05 \lfloor 35^\circ\ \Omega$.

260. The current in the circuit shown is
   (a) $(25 + j\,0)$ A
   (b) $(8.33 + j\,0)$ A
   (c) $(8.33 \lfloor 0)$ A
   (d) $(75 \lfloor 0)$ A.

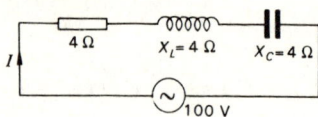

261. For a current of $(25 + j\,8)$ A to flow in the circuit shown, the applied voltage must be
   (a) $(175 + j\,56)$ V
   (b) $(107 - j\,76)$ V
   (c) $(99 - j\,132)$ V
   (d) $(25 + j\,8)$ V.

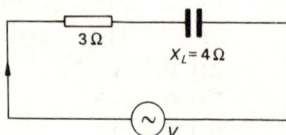

262. The impedance of the circuit shown can be represented by
   (a) $Z = R + j\,(X_L - X_C)$
   (b) $Z = R - j\,(X_L - X_C)$
   (c) $Z = R + j\,(X_L + X_C)$
   (d) $Z = R + j\,(\dfrac{X_L}{X_C})$.

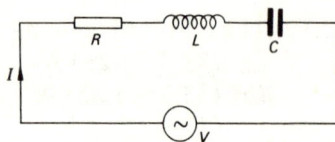

263. The impedance of the circuit shown is
   (a) $370 \lfloor 64^o23^1$
   (b) $277 \lfloor -64^o23^1$
   (c) $130 \lfloor 64^o23^1$
   (d) $872 \lfloor -64^o23^1$.

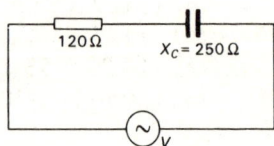

264. The current in the circuit below is given by
   (a) $I = \dfrac{V}{R}$
   (b) $I = \dfrac{V}{R + jX_L}$

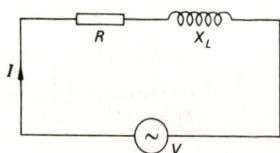

(c) $I = \dfrac{V}{R + jX_L}$

(d) $I = \dfrac{V}{R + jX_L}$.

265. The current in the circuit below is given by

(a) $I = \dfrac{V}{R}$

(b) $I = \dfrac{V}{R + jX_C}$

(c) $I = \dfrac{V}{R - jX_C}$

(d) $I = \dfrac{V}{\sqrt{R + jX_C}}$

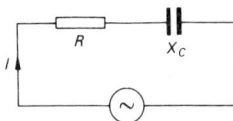

266. The voltage AB in the circuit below will be given by

(a) $V_{AB} = I\,(R - jX_L + jX_C)$
(b) $V_{AB} = I\,(R + jX_L - jX_C)$
(c) $V_{AB} = I\,(R^2 + j\,(X_L - X_C)$
(d) $V_{AB} = IR$.

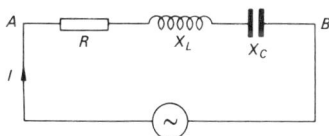

The significance of series LCR circuits at resonance

267. A series tuned circuit containing inductance, capacitance and resistance is resonant. If the value of the resistance is increased

(a) the circuit will still remain resonant but the current drawn from the supply will decrease
(b) the resonant frequency will decrease but the supply current will remain constant
(c) the resonant frequency will increase but the supply current will remain constant
(d) the circuit magnification will increase.

268. In the circuit shown below the impedance at resonance
     is
     (a) R
     (b) $\dfrac{L}{CR}$
     (c) infinite
     (d) zero.

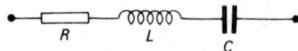

269. The resonant frequency of the circuit shown is approx-
     imately
     (a) 400 kHz
     (b) 4 MHz
     (c) 0.004 MHz
     (d) 400 Hz.

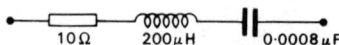

270. A series circuit has a resistance of 5 Ω, an inductance
     of 300 microhenries and a capacitance of 3 microfards.
     The impedance at resonance is
     (a) 5 Ω
     (b) 10 Ω
     (c) 20 Ω
     (d) 30 x 10^{-6} Ω.

271. An inductor of 100 μH and 5 Ω resistance is in series
     with a capacitor of 300 pF. A voltage of 10 V is applied
     at the resonant frequency; the current is
     (a) 150 μA
     (b) 15 μA
     (c) 2 A
     (d) 0.1 A.

272. At the resonant frequency of the circuit shown
     (a) V_L and V_C are in
         antiphase
     (b) V_L is at a minimum
     (c) V_R leads E by 90°
     (d) V_L lags I by 90°.

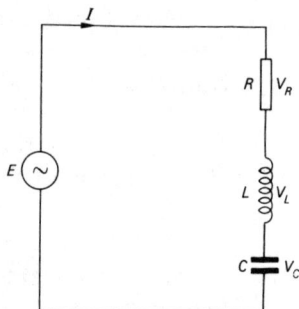

273. In the circuit shown Q_0 is

    (a) 2000
    (b) 25×10^{-4}
    (c) 100
    (d) 10.

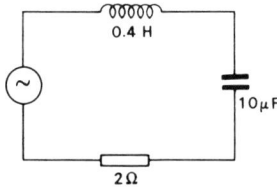

274. A series resonant circuit has a Q of 150. If the applied voltage at the resonant frequency is 16 volts, the voltage across the capacitor in the circuit
    (a) depends upon the ratio of the resistance to capacitance
    (b) is approximately 0.109 volts
    (c) is 2400 volts
    (d) equals the applied voltage.

275. In the circuit shown the power supplied at resonance is
    (a) 100 W
    (b) 20 W
    (c) 0 W
    (d) 1 W.

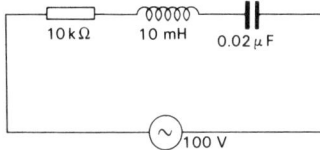

276. The resonant frequency of a series LCR circuit is 600 Hz. If the value of L is increased four times and the value of C increased nine times the new value of resonant frequency will be
    (a) 900 Hz
    (b) 100 Hz
    (c) 3600 Hz
    (d) 17 Hz.

277. In a series LCR circuit at resonance, if the frequency of the applied voltage is increased and the value of L, C and R are kept the same, the circuit will become
    (a) inductive
    (b) capacitive
    (c) resistive
    (d) resonant.

The use of phasor diagrams in the analysis of parallel R, L and C circuits

278. In the circuit shown the amplitude of the applied signal is kept constant but the frequency is increased
    (a) the impedance of the circuit increases
    (b) the current through R increases
    (c) the supply current increases
    (d) the phase angle increases.

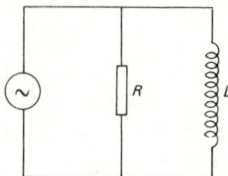

279. The magnitude of the supply current in the circuit shown is approximately
    (a) 12 A
    (b) 2 A
    (c) 4.8 A
    (d) 8.6 A.

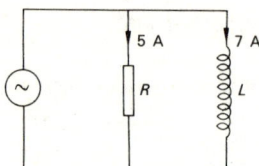

280. In the circuit shown, if the amplitude of the applied voltage remains constant but the frequency is reduced
    (a) the impedance of the circuit increases
    (b) the current through R increases
    (c) the current through L increases
    (d) the current through R decreases.

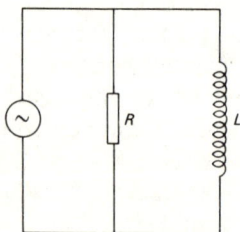

281. The impedance of the circuit below is
    (a) 35 Ω
    (b) 25 Ω
    (c) 12 Ω
    (d) 24 Ω.

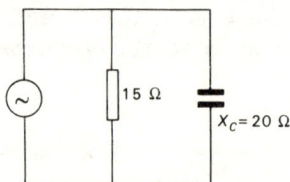

282. At the frequency at which $X_C = R = 1$ Ω, the impedance of the circuit shown is

    (a) 0.5 Ω
    (b) 0.707 Ω
    (c) 1 Ω
    (d) 1.414 Ω.

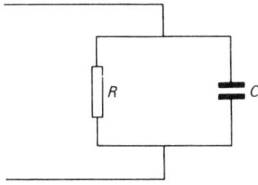

283. In a parallel circuit containing only a pure inductance L and a pure capitance C, $I_L = 12$ A and $I_C = 5$ A. The magnitude of the current taken from the supply is

    (a) 17 A
    (b) 13 A
    (c) 7 A
    (d) 5 A.

284. An inductance of 160 μH and a capacitor of 0.001 μF are connected in parallel across a supply of 100 V at 800 kHz. The effective reactance is

    (a) capacitive
    (b) inductive
    (c) resistive
    (d) capacitive and inductive.

285. In the circuit shown $I_R = 4$ mA, $I_C = 12$ mA, $I_L = 15$ mA. The magnitude of I is

    (a) 31 mA
    (b) 7 mA
    (c) 23 mA
    (d) 5 mA.

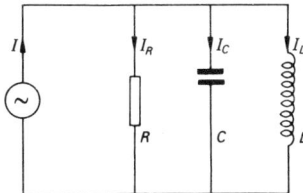

The significance of the parallel resonant condition of an R and L series circuit in parallel with C

286. At the resonant frequency of the circuit shown

    (a) I_L leads I by 90^0.

(b) I_C is a minimum
(c) I is in phase with E
(d) I_C lags I_L by 90°.

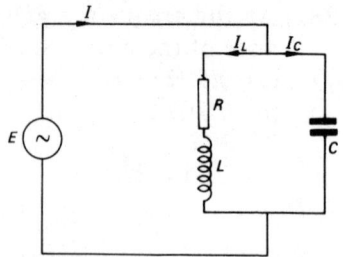

287. The circuit shown is at resonance. The impedance is:
(a) 200 MΩ
(b) 2 Ω
(c) 200 Ω
(d) 12.5 μΩ.

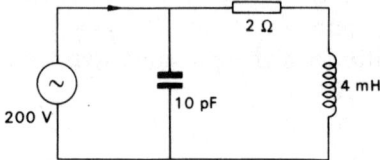

288. The supply current when the circuit shown is at resonance is
(a) 1 A
(b) 1 μA
(c) 100 A
(d) 400 A.

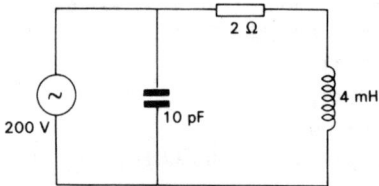

289. If the value of R in the circuit shown is increased and the supply voltage remains constant at the resonant frequency, the supply current will
(a) decrease
(b) lead the supply voltage
(c) lag the supply voltage
(d) increase.

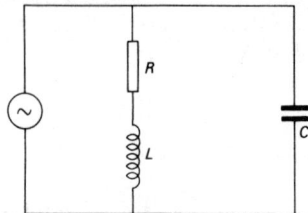

290. A loss free inductor L, a loss free capacitor C and a resistor R are connected in parallel across an alternating supply, the frequency of which is adjusted so that L

and C resonate. L and C are then removed. It will be found that

(a) the current drawn from the supply increases

(b) the current drawn from the supply decreases

(c) the supply current remains unaltered in magnitude and phase

(d) the magnitude of the supply current is unchanged but the phase angle alters.

291. In the parallel circuit shown the resonant frequency is f_o, and the current magnification is Q_o. If one of the capacitors becomes short-circuited, then

(a) f_o and Q_o both increase

(b) f_o increases, Q_o decreases

(c) f_o decreases, Q_o increases

(d) f_o decreases and Q_o decreases.

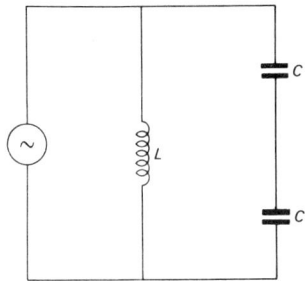

292. In the circuit shown, if L and R remain fixed the effect of increasing C will be to

(a) increase the resonant frequency

(b) reduce the dynamic impedance

(c) increase the Q factor

(d) increase the dynamic impedance.

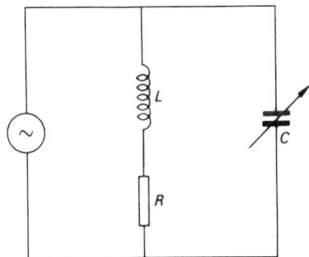

293. In a parallel tuned circuit

(a) the circuit impedance is a maximum at resonance

(b) voltage magnification takes place

(c) impedance at resonance is low

(d) the current taken from the supply is greater than the circulating currents.

294. The dynamic impedance of a parallel circuit at resonance is 25 kΩ. If $L = 10$ mH and $R = 4$ Ω, the value of C is
    (a) 250 pF
    (b) 62.5 pF
    (c) 0.1 μF
    (d) 5 μF.

295. An inductor is connected in parallel with a capacitor. If the inductance is doubled, and the capacitance is halved, the impedance at resonance will be
    (a) doubled
    (b) reduced by a factor of four
    (c) multiplied by a factor of four
    (d) unaltered.

296. A parallel LCR circuit when operated at a frequency above its resonant frequency is
    (a) inductive
    (b) resistive
    (c) capacitive
    (d) inductive and resistive.

297. The graph of impedance against frequency for a parallel LCR circuit is
    (a) A
    (b) B
    (c) C
    (d) D.

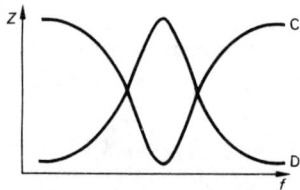

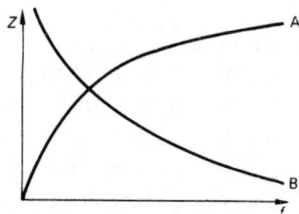

The basic construction and principles of action of a single phase transformer

298. The core of a transformer is laminated in order to:
   (a) decrease the inductance of the windings
   (b) prevent hysteresis loss
   (c) reduce the effects of eddy currents
   (d) prevent eddy currents from occuring.

299. In the circuit shown, the current in part *AB* of the winding is
   (a) 1.7 A
   (b) 0.7 A
   (c) 1.2 A
   (d) 0.5 A.

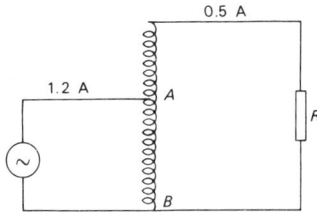

300. Assuming that the total flux in a transformer links with every turn, which of the following relationships is true?
   (a) $I_p N_p = I_s N_s$

   (b) $I_p I_s = N_p N_s$

   (c) $\dfrac{N_s}{N_p} = \sqrt{I_p}$

   (d) $\dfrac{N_s^2}{N_p} = I_p I_s.$

301. The power input to a mains transformer is 100 W, the primary current is 1 A and the secondary voltage is 10 V. Assuming no losses in the transformer, the turns ratio is
   (a) 10:1 step down
   (b) 10:1 step up
   (c) 100:1 step down
   (d) 100:1 step up.

302. A step up transformer has a turns ratio of 5. If the output current is 2.5 A, the input current is
   (a) 0.5 A
   (b) 1.5 A
   (c) 1.25 A
   (d) 12.5 A.

303. A loss free transformer has a resistor connected across its output terminals. If the value of the resistor is reduced
   (a) the output voltage will fall
   (b) the input current will fall
   (c) the input current will increase
   (d) the input voltage will fall.

304. If the frequency of the input signal to a transformer is increased the e.m.f. induced across the primary winding will
   (a) increase
   (b) decrease
   (c) vary as the square of the frequency
   (d) remain unchanged.

305. The phase angle between the primary current and primary voltage of a loss free transformer when operating with a secondary load will normally be
   (a) primary current leading primary voltage by 90°
   (b) primary current lagging primary voltage by 90°
   (c) primary current lagging primary voltage by less than 90°
   (d) primary current lagging primary voltage by more than 90°.

306. A voltage step-up transformer has
   (a) a larger current in the secondary than the primary
   (b) a larger power output in the secondary than the primary
   (c) a smaller current in the secondary than the primary
   (d) more turns in the primary than the secondary.

307. Iron losses in a transformer are due to
  (a) the resistance of the primary and secondary windings
  (b) eddy currents only
  (c) flux leakage
  (d) both eddy current and hysteresis losses.

308. An advantage of an autotransformer is that
  (a) the primary and secondary windings are not separated hence iron losses are reduced
  (b) copper losses are reduced when the turns ratio is small
  (c) it gives a high step-up ratio
  (d) it reduces inter-turn capacitance.

309. The e.m.f. equation of a transformer of primary turns N_1, maximum flux density Bm, magnetic area of core a, operating at frequency f is given by
  (a) $E_1 = \dfrac{N_1\, Bm\, f}{a}$ volts
  (b) $E_1 = 4.44\, \dfrac{N_1\, Bm\, f}{a}$ volts
  (c) $E_1 = 4.44\, N_1\, Bm\, f\, a$ volts
  (d) $E_1 = 1.11\, N_1\, Bm\, f\, a$ volts.

The construction and action of a.c. motors and generators

310. The frequency of the output voltage of an alternator depends upon the
  (a) rotor speed only
  (b) rotor speed and the number of poles
  (c) number of poles only
  (d) the magnetic field strength.

311. The speed of a synchronous motor depends upon:
  (a) field current
  (b) slipspeed
  (c) hysteresis loss
  (d) supply frequency.

312. The slip speed of an induction motor may be defined as
    (a) rotor speed − synchronous speed
    (b) synchronous speed − rotor speed
    (c) rotor speed + synchronous speed
    (d) number of pairs of poles × 60 ÷ frequency.

313. A three-phase alternator rotates at a speed of 3000 r.p.m. and has eight pairs of poles. The supply frequency is
    (a) 2.4 Hz
    (b) 50 Hz
    (c) 375 Hz
    (d) 400 Hz.

314. The speed of rotation of the rotor in a synchronous motor is governed by the
    (a) voltage applied to the field windings
    (b) frequency of the supply voltage
    (c) current in field windings
    (d) the load put on the motor

315. An alternator has eight poles, set in four pairs, on its rotor which revolves at 3000 r.p.m. The frequency of the a.c. generated will be
    (a) 200 Hz
    (b) 400 Hz
    (c) 750 Hz
    (d) 12 Hz.

316. The slip speed of an induction motor depends upon:
    (a) mechanical load
    (b) supply voltage
    (c) armature current
    (d) eddy currents.

317. An alternating current generator has two pairs of poles. If the armature is rotated at 3000 r.p.m. the output frequency will be
    (a) 1500 Hz   (c) 100 Hz
    (b) 200 Hz   (d) 50 Hz.

318. The starting torque of a simple squirrel cage motor is
     (a) low
     (b) high
     (c) decreases as rotor current rises
     (d) increases as rotor current rises.

319. The slip speed of an induction motor
     (a) is 100 per cent until the rotor moves and then falls
         to a low value
     (b) is 0 until the rotor moves and then rises to 100 per
         cent
     (c) is 100 per cent until the rotor moves and then
         decreases slightly
     (d) is 0 until the rotor moves and then rises slightly.

320. A six pole induction motor when fed from a 50 Hz
     supply experiences a 7 per cent slip. The rotor speed
     will, therefore, be
     (a) 15 rev./s
     (b) 15.5 rev./s
     (c) 16 rev./s
     (d) 16.5 rev./s.

The basic construction and action of d.c. motors and generators

321. The speed of a d.c. motor may be increased by
     (a) increasing the field current
     (b) decreasing the applied voltage
     (c) decreasing the field current
     (d) increasing the armature current.

322. The armature resistance of a d.c. motor is 0.5 Ω, the
     applied voltage is 100 V and the back e.m.f. is 98 V
     at full speed. The armature current is
     (a) 4 A   (c) 200 A
     (b) 2 A   (d) 196 A.

323. A starter resistor is necessary in large d.c. motors to
  (a) increase the armature starting current
  (b) reduce the armature starting current
  (c) reduce the field current
  (d) increase the back e.m.f.

324. If a resistance is inserted in series with the field winding
of a shunt wound motor, the effect is to
  (a) reduce the speed of the motor
  (b) increase the magnetic field
  (c) increase the speed of the motor
  (d) decrease the armature current.

325. The effect of increasing the field current of a d.c.
generator is to
  (a) increase the speed of the generator
  (b) decrease the armature current
  (c) increase the output voltage
  (d) reduce the magnetic field strength.

326. A back e.m.f. is produced in a d.c. motor when rotating
because
  (a) eddy currents are produced in the core
  (b) the armature conductors are cutting lines of
    magnetic flux
  (c) additional current is induced in the field coils
  (d) a voltage drop occurs in the armature windings.

327. Reversing the polarity of the input voltage to a series
d.c. motor will
  (a) have no effect
  (b) increase the armature current
  (c) reverse the direction of rotation
  (d) reduce starting torque.

328. The output voltage of a d.c. generator can be increased
by
  (a) decreasing the speed of rotation of the armature
  (b) increasing the magnetic flux density in the air gap
  (c) decreasing the number of turns of the field winding

(d) ensuring that the air gap between armature and field poles is sufficiently wide.

329. The direction of rotation of a d.c. motor can be reversed by reversing the
(a) current in the field winding only
(b) connections in both field and armature windings
(c) supply voltage
(d) current in the armature and field windings.

330. A starter resistor is often used with d.c. motors. This resistor
(a) limits the field current to a safe value on starting;
(b) controls the speed of the machine
(c) limits the armature current to a safe value on starting
(d) prevents the field current from flowing through the armature and damaging it.

331. The effective armature voltage of a d.c. motor is equal to the
(a) armature supply voltage
(b) armature supply voltage − the back e.m.f.
(c) back e.m.f.
(d) armature supply voltage + the back e.m.f.

332. The correct curve for a shunt wound d.c. generator is
(a) A
(b) D
(c) C
(d) B.

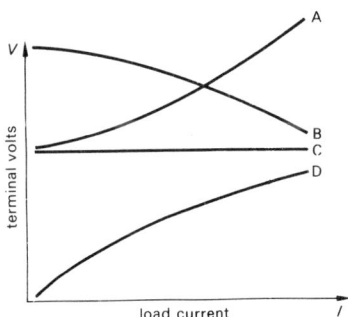

333. In d.c. generators iron losses are made up of
     (a) hysteresis, eddy current and copper losses
     (b) hysteresis and eddy current losses
     (c) hysteresis and friction losses
     (d) hysteresis, eddy current losses, plus contact resistance losses of the brushes.

334. The voltage/speed characteristic curves for a simple d.c. generator are shown for three different values of field current. It is true to say that
     (a) If_3 is greater than If_2 and If_1
     (b) If_1 is greater than If_2 and If_3
     (c) If_3 is less than If_1 and If_2
     (d) If_2 is greater than If_3 and less than If_1.

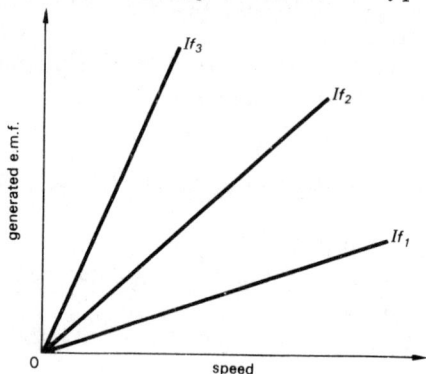

335. The graph shown represents the speed/torque characteristic of a
     (a) shunt motor
     (b) differential compound motor
     (c) series motor
     (d) induction motor.

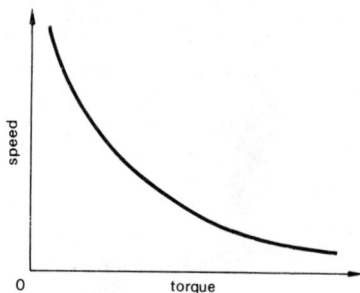

336. In the graphs of speed/torque shown below the characteristics of a d.c. shunt motor are most closely represented by
    (a) A
    (b) B
    (c) C
    (d) D.

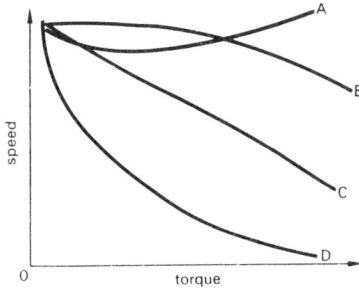

337. The graph shows the output voltage/load current characteristic of a separately excited d.c. generator. The drop in output voltage *AB* is most likely to be caused by
    (a) voltage drop in the armature
    (b) armature reaction
    (c) voltage drop in the armature + armature reaction
    (d) voltage drop in the armature − armature reaction.

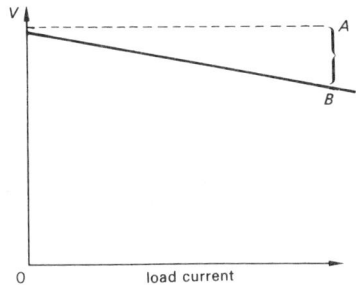

338. The output voltage/load current graph shown is that of a
    (a) separately excited d.c. generator
    (b) series wound d.c. generator
    (c) shunt wound d.c. generator
    (d) shunt wound d.c. motor.

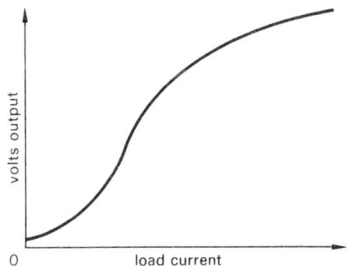

The basic principles of three-phase systems and solving problems involving balanced systems only

339. In a three phase delta connected system the relationships between line and phase values of V and I are

(a) $V_L = V_{pb}$ and $I_L = \sqrt{3I_{pb}}$

(b) $V_L = \sqrt{3V_{pb}}$ and $I_L = I_{pb}$

(c) $V_L = V_{pb}$ and $I_L = \dfrac{I_{pb}}{\sqrt{3}}$

(d) $V_L = \dfrac{V_{pb}}{\sqrt{3}}$ and $I_L = I_{pb}$.

340. A three phase alternator has its stator windings connected in delta. If the phase voltage and current are 230 V and 100 A respectively, the line voltage and current will be

(a) 230 V and $100\sqrt{3}$ A

(b) $230\sqrt{3}$ V and 100 A

(c) $\dfrac{230\ \text{V}}{\sqrt{3}}$ and 100 A

(d) 230 V and $\dfrac{100\ \text{A}}{\sqrt{3}}$.

341. The line voltage of a four wire three-phase star connected system is 11 kV. The phase voltage will be

(a) 7778 V

(b) 11 kV

(c) 19.05 kV

(d) 6351 V.

342. The line current of a four wire three-phase star connected system is 5 A. The phase current will be

(a) 20 A   (c) 10 A

(b) 15 A,   (d) 5 A.

343. The phase voltage of a four wire three-phase star connected system is 120 V. The line voltage will be approximately
   (a) 480 V   (c) 208 V
   (b) 360 V   (d) 120 V.

344. In a delta connected three-phase system with balanced loads the phase voltage is 240 V. The line voltage will be
   (a) 240 V   (c) 440 V
   (b) 340 V   (d) 720 V.

345. In a delta connected three-phase system with balanced loads the line current is 1.73 A. The phase current will be
   (a) 1.73 A   (c) 0.707 A
   (b) 1.0 A   (d) 1.414 A.

346. In the four wire star connected three-phase system shown the neutral conductor would be
   (a) A
   (b) B
   (c) C
   (d) D.

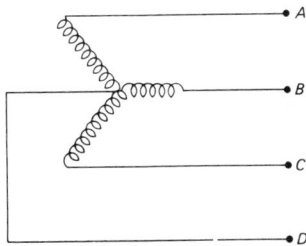

347. In the four wire star connected three-phase system shown the voltage between A and B is the
   (a) phase voltage
   (b) line voltage
   (c) sum of two line voltages
   (d) difference of two line voltages.

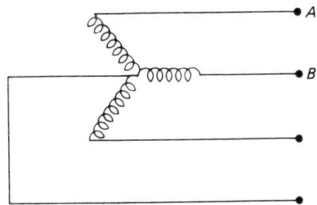

348. In a three-phase four wire star connected system with balanced loads, the phase currents are each 3 A. An ammeter placed in the common return conductor will indicate
   (a) 0 A
   (b) 3 A
   (c) 6 A
   (d) 9 A.

The significance of hysteresis in magnetic circuits

349. The intercept of the hysteresis curve on the B axis represents
   (a) remanent flux density
   (b) coercive force
   (c) reluctance
   (d) reactance.

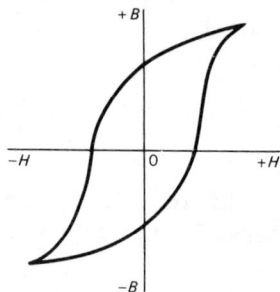

350. A magnetic specimen has a flux density at saturation of 1.2 tesla. It is then demagnetised by the application of a magnetic field strength of 200 A/m. The coercivity of the specimen will be
   (a) 1.2 tesla
   (b) 240 A/m
   (c) 200 A/m
   (d) 166 A/m.

351. The figure shows the hysteresis curves for two different magnetic specimens, X and Y. The energy loss of material X is
   (a) less than Y
   (b) greater than Y

(c) proportional to B^2
(d) proportional to H^2.

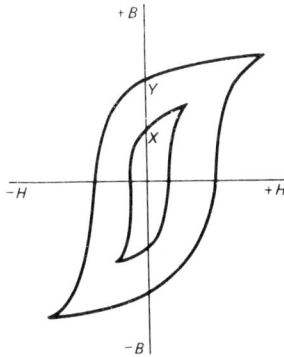

352. The hysteresis loss in a transformer can be reduced by
    (a) increasing the transformation ratio
    (b) using mumetal as a core material
    (c) using soft iron as a core material
    (d) increasing the size of the transformer core.

353. The figure shows the hysteresis curves for two magnetic
    materials, X and Y. The curves show that Y would be
    a more suitable material than X for use as a transformer
    core because
    (a) it has a lower
        remanence
    (b) it has a higher
        coercivity
    (c) it has a smaller
        hysteresis loss
    (d) it has a higher
        hysteresis loss.

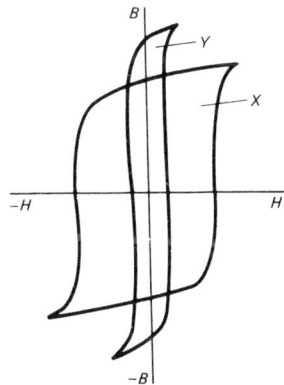

The concept of energy stored in an electric field between charged parallel plates

354. A fully charged 50 μF capacitor has a potential difference of 200 V across its plates. The electrostatic energy stored will be
(a) 10 J
(b) 1 J
(c) 0.1 J
(d) 1 mJ.

355. The plates of a parallel plate capacitor are set 10 mm apart in a vacuum. Given that $\Sigma_0 = 8.85 \times 10^{-12}$ F/m and the voltage across the plates is 1000 V, the energy per cubic metre will be
(a) 8.85×10^{-2} J
(b) 8.85 J
(c) 4.42×10^{-2} J
(d) 4.42 J.

356. A 200 μF capacitor is connected to a d.c. supply of V volts until fully charged. If the energy stored is 1 joule the voltage V is
(a) 1000 V
(b) 400 V
(c) 200 V
(d) 100 V.

357. Electric flux density is measured in
(a) volts per metre
(b) volts per metre^2
(c) coulombs per metre
(d) coulombs per metre^2.

The construction and principles of the wattmeter

358. An electrodynamic wattmeter is connected to measure the power in a load. The fixed and moving coils are connected
(a) in series with each other, and together carry the load current

(b) across the supply and in series with the load respectively

(c) in series with the load and across the supply respectively

(d) in parallel with each other so as to share the load current.

359. In an ideal electrodynamic wattmeter, the fixed and moving coil circuits have
(a) very high and very low resistances respectively
(b) both very low resistances
(c) both very high resistances
(d) very low and very high resistances respectively.

360. The damping torque in an electrodynamic wattmeter is produced by
(a) eddy currents in an aluminium former carrying the current coil
(b) eddy currents in an aluminium former carrying the voltage coil
(c) pneumatic action due to air friction
(d) hairsprings attached to coils.

Answers to Questions

Chapter 1: Level I

1. (b)	21. (b)	41. (b)	61. (c)	81. (b)	101. (b)
2. (d)	22. (b)	42. (c)	62. (b)	82. (b)	102. (c)
3. (a)	23. (a)	43. (d)	63. (d)	83. (a)	103. (a)
4. (c)	24. (a)	44. (c)	64. (d)	84. (d)	104. (d)
5. (b)	25. (a)	45. (b)	65. (b)	85. (c)	105. (d)
6. (d)	26. (b)	46. (d)	66. (c)	86. (b)	106. (b)
7. (c)	27. (c)	47. (d)	67. (c)	87. (b)	107. (c)
8. (d)	28. (d)	48. (b)	68. (c)	88. (c)	108. (b)
9. (b)	29. (b)	49. (c)	69. (a)	89. (c)	109. (b)
10. (b)	30. (d)	50. (a)	70. (d)	90. (c)	110. (c)
11. (a)	31. (b)	51. (a)	71. (b)	91. (a)	111. (a)
12. (b)	32. (c)	52. (b)	72. (a)	92. (c)	112. (d)
13. (b)	33. (c)	53. (a)	73. (d)	93. (d)	113. (c)
14. (b)	34. (a)	54. (b)	74. (d)	94. (c)	114. (b)
15. (d)	35. (b)	55. (d)	75. (d)	95. (a)	115. (c)
16. (a)	36. (c)	56. (d)	76. (d)	96. (a)	116. (b)
17. (a)	37. (c)	57. (b)	77. (b)	97. (b)	117. (b)
18. (a)	38. (b)	58. (c)	78. (c)	98. (a)	118. (a)
19. (a)	39. (b)	59. (d)	79. (a)	99. (d)	119. (c)
20. (c)	40. (d)	60. (a)	80. (c)	100. (a)	120. (c)

Chapter 2: Level II

121. (b)	141. (a)	161. (c)	181. (d)	201. (b)	221. (d)
122. (d)	142. (a)	162. (c)	182. (c)	202. (d)	222. (b)
123. (a)	143. (b)	163. (c)	183. (b)	203. (d)	223. (d)
124. (d)	144. (b)	164. (d)	184. (b)	204. (c)	224. (c)
125. (b)	145. (a)	165. (a)	185. (b)	205. (a)	225. (a)
126. (d)	146. (a)	166. (d)	186. (a)	206. (c)	226. (c)
127. (a)	147. (d)	167. (a)	187. (d)	207. (c)	227. (d)
128. (c)	148. (a)	168. (d)	188. (a)	208. (d)	228. (b)
129. (b)	149. (a)	169. (b)	189. (c)	209. (c)	229. (b)

130. (a)	150. (b)	170. (a)	190. (a)	210. (c)	230. (a)
131. (b)	151. (a)	171. (d)	191. (c)	211. (b)	231. (a)
132. (d)	152. (d)	172. (b)	192. (d)	212. (b)	232. (c)
133. (c)	153. (a)	173. (c)	193. (b)	213. (c)	233. (b)
134. (a)	154. (b)	174. (d)	194. (b)	214. (c)	234. (d)
135. (a)	155. (c)	175. (c)	195. (b)	215. (b)	235. (c)
136. (b)	156. (c)	176. (d)	196. (c)	216. (b)	236. (d)
137. (d)	157. (b)	177. (c)	197. (b)	217. (a)	237. (c)
138. (a)	158. (a)	178. (c)	198. (d)	218. (d)	238. (b)
139. (d)	159. (c)	179. (b)	199. (d)	219. (a)	239. (d)
140. (a)	160. (b)	180. (d)	200. (b)	220. (d)	240. (a)

Chapter 3: Level III

241. (c)	261. (b)	281. (c)	301. (a)	321. (c)	341. (d)
242. (a)	262. (a)	282. (b)	302. (d)	322. (a)	342. (d)
243. (c)	263. (b)	283. (c)	303. (c)	323. (b)	343. (c)
244. (c)	264. (b)	284. (a)	304. (a)	324. (c)	344. (a)
245. (a)	265. (c)	285. (d)	305. (c)	325. (c)	345. (b)
246. (c)	266. (b)	286. (c)	306. (c)	326. (b)	346. (d)
247. (a)	267. (a)	287. (a)	307. (d)	327. (c)	347. (b)
248. (b)	268. (a)	288. (b)	308. (b)	328. (b)	348. (a)
249. (b)	269. (a)	289. (d)	309. (c)	329. (a)	349. (a)
250. (c)	270. (a)	290. (b)	310. (b)	330. (c)	350. (c)
251. (c)	271. (c)	291. (d)	311. (d)	331. (b)	351. (a)
252. (a)	272. (a)	292. (b)	312. (b)	332. (d)	352. (b)
253. (d)	273. (c)	293. (a)	313. (d)	333. (b)	353. (c)
254. (c)	274. (c)	294. (c)	314. (b)	334. (a)	354. (b)
255. (b)	275. (d)	295. (c)	315. (a)	335. (c)	355. (c)
256. (a)	276. (b)	296. (c)	316. (a)	336. (b)	356. (d)
257. (c)	277. (a)	297. (d)	317. (c)	337. (a)	357. (d)
258. (d)	278. (a)	298. (c)	318. (a)	338. (b)	358. (c)
259. (b)	279. (d)	299. (b)	319. (a)	339. (a)	359. (d)
260. (a)	280. (c)	300. (a)	320. (b)	340. (a)	360. (c)

Bibliography

Anastasi, A.,	*Psychological Testing*, Macmillan
Gronlund, N.E.,	*Measurement and Evaluation in Teaching*, Macmillan
Ebel, R.L.,	*Essentials of Educational Measurement*, Prentice Hall
Thyne, J.M.,	*Principles of Examining*, University of London Press.
Objective Testing,	City & Guilds of London (1970)
Adkins, D.C.,	*Test Construction*, C.E. Merrill Books
Kelley, T.L.,	*Jr. of Ed. Psychology* (1939) p.17–24
Anderson, S.B., & Katz, M.,	*Test Construction*, C.E. Merrill Books
Gerberich, R.,	*Specimen Objective Test Items*, David McKay Co. Inc. (New York)
Jessup, Knight & Haslett,	*Manual of Trade Proficiency Testing*, London M.O.D. (Air)
Thorndike, R.L.,	*Educational Measurement* Washington D.C., American Council on Education, 1971
Vernon, P.E.,	*Intelligence and Attainment Tests*, University of London Press
Wood, D.A.,	*Test Construction*, C.E. Merrill Books

Index

For a full list of titles and prices, write
for the FREE TEC/BEC leaflet and/or the Macdonald
& Evans Business Studies catalogue, available from
Department TB1, Macdonald & Evans Ltd.,
Estover, Plymouth PL6 7PZ